To Utopia and Back:
The Search for Life in the Solar System

To Utopia and Back: The Search for Life in the Solar System

Norman H. Horowitz
California Institute of Technology

W. H. Freeman and Company
New York

Library of Congress Cataloging-in-Publication Data

Horowitz, Norman Harold, 1915–
  To utopia and back.

  Bibliography: p.
  Includes index.
  1. Life on other planets. 2. Mars (Planet)—
Exploration. 3. Viking Mars Program. I. Title.
QB54.H654 1986   574.999'2   85-27351
ISBN 0-7167-1765-4
ISBN 0-7167-1766-2 (pbk.)

Printed in the United States of America

1 2 3 4 5 6 7 8 9 0   MP   4 3 2 1 0 8 9 8 7 6

TO PEARL

Fan and critic, stylist, grammarian, motivator—
she would have made it better

Contents

CHAPTER FIVE
Mars: Myth and Reality   76

CHAPTER SIX
**The Viking Mission: Water, Life,
and the Martian Desert**   101

Preface

The exploration of the planet Mars in the 1960s and 1970s, culminating in the landings of two instrumented Viking spacecraft in 1976, ended a long dream of Western civilization. Since these explorations obtained a negative answer to their most interesting question, that of Martian life, they are doubtless regarded by many as a failure. In actuality, however, they were a dazzling success. A series of six beautifully designed and operated space missions transformed Mars in the course of a decade from a mysterious planet with a long, romantic history into one of the most familiar places in the solar system. To an academic scientist caught up in these investigations, the entire Martian enterprise was and remains as extraordinary as any first exploration of a new world must seem. The range of scientific interests it engaged, the sophistication of its instruments, the size and complexity (and effectiveness) of its organization, the sense of excitement and historic importance it inspired in participants and audience—all these give the exploration of Mars a singular place in the history of our time.

Failure to find life at either of the Viking landing sites, together with persuasive evidence that other Martian locales are no more hospitable than the two sampled, argues strongly that terrestrial life is alone in the solar system, a conclusion of more than passing interest. For many who had followed the course of Martian explorations, starting with the flyby of *Mariner 4* in 1965, the final outcome was not surprising. For many others, however, probably including much of the general public, it was a disappointment that may not be fully accepted even yet.

The idea of an inhabited Mars occupies a special place in our culture, and this idea had powerfully influenced the pace and style of the Martian explorations. It explains, to begin with, the remarkable fact that the entire series of Martian expeditions was completed within 20 years of the launch of *Sputnik 1* in October, 1957, the inaugural event of the space age. In addition, the belief that Mars was simply a more hostile version of the earth—persisted in despite abundant evidence to the contrary—strongly influenced the selection of biological experiments for the Viking mission, and it underlay an international agreement that imposed a stringent quarantine policy on all Mars-bound spacecraft. The power of the Martian dream to distort the judgment of normally sober scientists is also visible in illusory observations of the planet that were still being reported in very recent times. The record of these claims and their eventual correction, chiefly by spacecraft during the period between 1965 and 1976, forms one of the most extraordinary chapters in the history of modern science.

My purpose in writing this book is to present an account of these matters as seen by a participant in language that is comprehensible to nonscientists. The like of the Mars project, with its unique mixture of legend, scientific incentive, technological capabilities, and public enthusiasm, will not be seen again soon, if only because there is not another object in the solar system with the irresistible allure of the old Mars. I shall also try to explain for the general reader the scientific background of the search for extraterrestrial life and for possible life-supporting environments, including the choice of Mars as the only plausible target in the solar system, and I hope to make clear why the search can now be considered closed.

I am happy to acknowledge valuable comments made on various chapters of the manuscript by the following people: Elizabeth Bertani, J. B. Farmer, Jesse Greenstein, David Horowitz, Jerry Hubbard, Andrew Ingersoll, Robert Leighton, Lynn Margulis, Stanley Miller, Bruce Murray, Leslie Orgel, Maarten Schmidt, and the late Ralph Robin. In addition, I was enlightened on a number of points by Benton Clark, John Edmond, Clifford Moran, Conway Snyder, David Stevenson, Steven Vogel, and Yuk Yung. My thanks go to all these friends. None of them is in any way responsible for whatever deficiencies remain in this account, and the views expressed here are mine alone.

I owe a special word of appreciation to Hardy Martel for his assistance on the occasions when my word processor would respond only to someone who could speak to its inner soul, I am obliged to Jenijoy La Belle for calling my attention to the poem by Theodore Roethke from which the lines that head Chapter 3 were taken, and I thank Cheryl Kupper for her skillful editing and Jurrie Van Der Woude for assistance with the illustrations.

N.H.H.

To Utopia and Back:
The Search for Life in the Solar System

CHAPTER ONE

"What Is Life?"

It is not very long since genetics and biochemistry were quite separate sciences, each seeking . . . a key to unlock the mystery of life. The biochemists found enzymes, the geneticists genes.

William Hayes, *The Genetics of Bacteria and Their Viruses* (1968)

The year is 1958. The age of space exploration has arrived with a rush, and you have been invited to assist in planning a search for life on the planet Mars. You accept—who could resist?—and soon find yourself thinking about looking for life on another world. How would one recognize an object on another planet as alive? What do we mean when we say that something is "living," anyway? Questions such as these have occupied philosophers as far back as Aristotle.

Ordinarily it is not hard to decide whether a thing is alive or not. We associate a variety of appearances and behaviors with the living state, and their absence usually leads to the correct conclusion that an object is nonliving. In especially interesting or important cases —whether or not seeds found in an Egyptian tomb are alive, whether or not a human being is dead—the results of certain tests settle the question. But are such criteria adequate for identifying life on Mars? It would not seem so. Our concept of "life" must be broad enough to let us recognize it in any guise and precise enough to prevent our finding it where it does not exist. To attain such a concept, we must delve into the fundamental nature of living matter, as modern biology has revealed it.

LIFE AND THE GENETIC MECHANISM

Living systems differ from inanimate ones in two general ways. First, the simplest living organism is enormously more complex in its composition and organization than anything of comparable size in inanimate nature. Even today, we do not fully know the structure of any cell: new cellular components continue to be discovered, and the end is not in sight. And this does not take into account the intricacies of organization found in multicellular creatures, where vast populations of cells specialized for different functions are constrained to act in a coordinated and mutually beneficial manner. This brings us to the second distinctive feature of living things: their design and construction *appear* to result from a conscious purpose, that purpose being to ensure the organism's survival. In the past, consideration of these unique properties led to the belief that living organisms embody a "vital force"—a mysterious, nonphysical principle that endows them with special attributes and separates them from the inanimate world by an unbridgeable gap. Notions such as this are no longer tenable. We now know that living matter does not differ in a truly fundamental way from nonliving matter. Living things are composed of atoms and molecules and nothing else. What distinguishes them from the rest of the universe is the way their atoms are put together: life is a manifestation of certain molecular combinations.

Only two kinds of molecules underlie the essential phenomena associated with life as we know it on our planet: proteins and nucleic acids. Proteins form enzymes, the highly efficient and versatile catalysts[1] that bring about the multifarious chemical reactions that occur in living systems. Chemical changes of one kind or an-

---

[1] See the Glossary for this and other technical terms.

other characterize all the activities of living things: the digestion of food, the formation of new cells and cell components, the contraction of muscle, and the transmission of nerve impulses, to mention only a few, result from chemical transformations in which molecules of one kind change into molecules of a different kind. These specific transformations, which represent only a tiny fraction of all the reactions possible with the substances available, are selected and initiated by enzymes, and thus enzymes determine the direction and outcome of the complex reaction networks called "metabolism" that characterize the living state.

Living cells produce proteins that have other functions as well. Nonenzymatic proteins include hemoglobin, insulin, and antibodies. The most abundant protein produced by mammals is collagen, a nonenzyme that has a structural role in bone, skin, and teeth.

Nucleic acids have quite a different function. They form the genes, the bearers of the genetic heritage, in all species. Both kinds of nucleic acid, deoxyribonucleic acid (DNA) and ribonucleic acid (RNA), are found in all cells. Although they are similar in chemical structure, DNA plays the genetic role in known organisms, except for some viruses. The genetic heritage consists of information, apparently all of it concerned with the production of protein molecules—their chemical structure, their time of appearance, and their rate of synthesis.

Both nucleic acids and proteins are large molecules composed of linear arrays of small subunits, or building blocks. The building blocks of nucleic acids are called nucleotides, four different kinds of which make up DNA and RNA (see Figure 1-1 for their structures). Genetic information is encoded in sequences of nucleotides, just as the information on a printed page is encoded in sequences of letters. The building blocks of proteins are amino acids. A great many amino acids exist in nature, but just 20—the same 20 in all known species—are used to build proteins (see Figure 1-2 for their structures).

An important characteristic of amino acids is their "optical isomerism." All the amino acids of proteins except the simplest one, glycine, can exist in two forms that are related to each other as the left and right hands are related—that is, they are mirror images, as

DNA

shown in Figure 1-3. The two optical isomers are identical in their chemical properties, but because they are not superimposable—a right-handed glove does not fit on the left hand—they cannot substitute for each other in the construction of protein molecules or in any other three-dimensional relation. Interestingly, the amino acids of the proteins of all known species are uniformly left-handed, or L (levo), isomers. In principle, all the amino acids in the living world could be right-handed, or D (dextro), isomers, and this world would function in the same way as the one we know. The fact

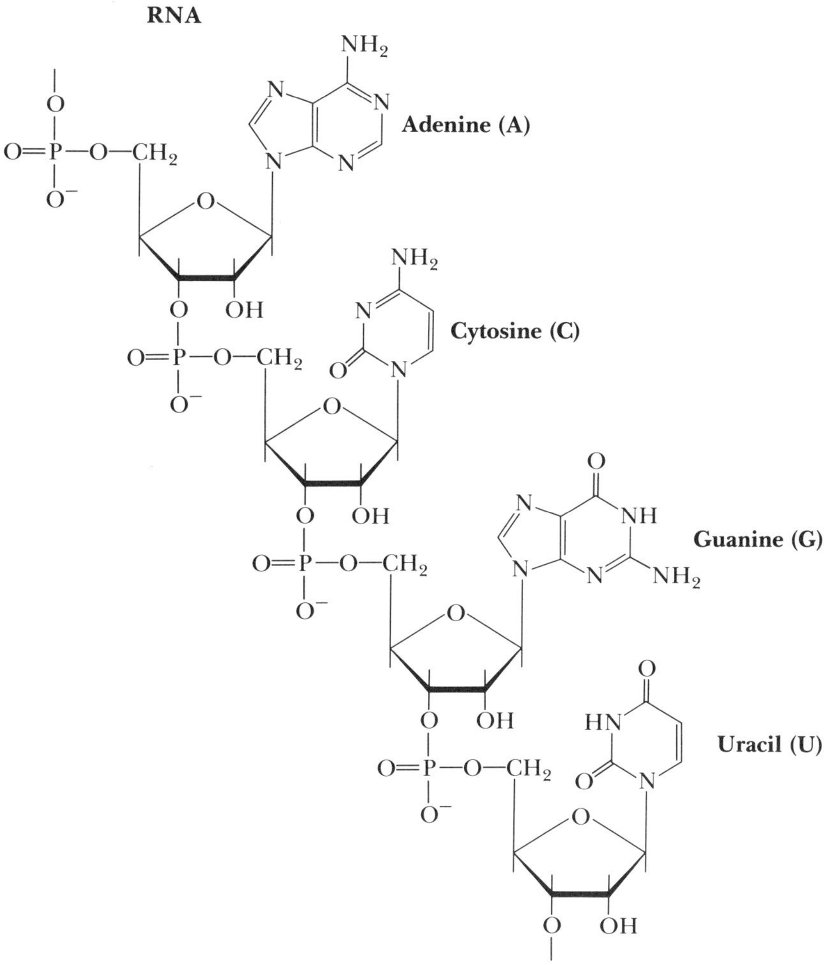

RNA

Figure 1-1   The four nucleotides of DNA and RNA joined together to make short nucleic acid segments. Each nucleotide consists of a nitrogen-containing base (its name and abbreviation are shown) attached to a five-carbon sugar (ribose in RNA, deoxyribose in DNA), and this in turn is attached to a molecule of phosphoric acid. Phosphoric acid forms the link between nucleotides in nucleic acid chains.

that we use L rather than D amino acids is probably a matter of chance, an accident of history. On another planet where amino

Glycine

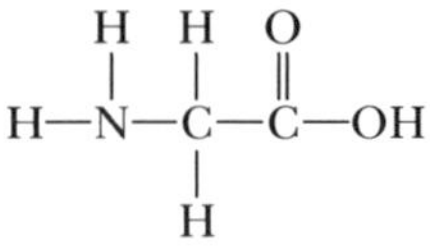

Alanine

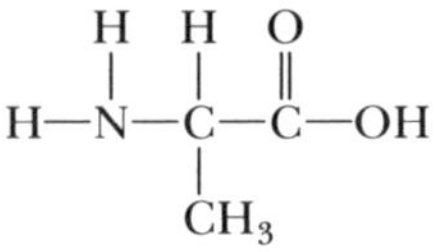

Valine

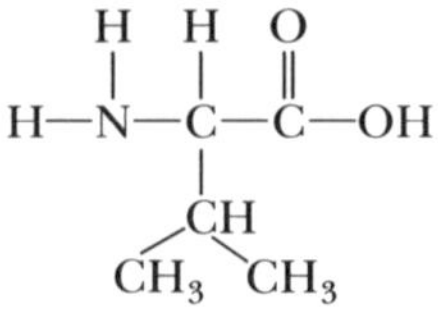

Leucine

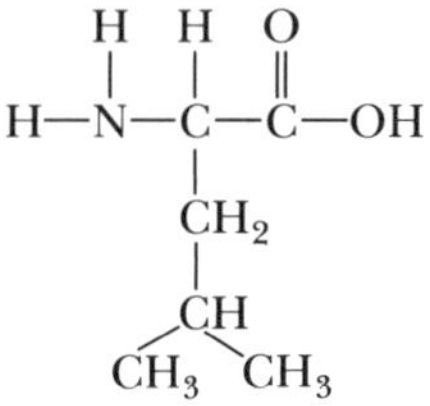

Isoleucine

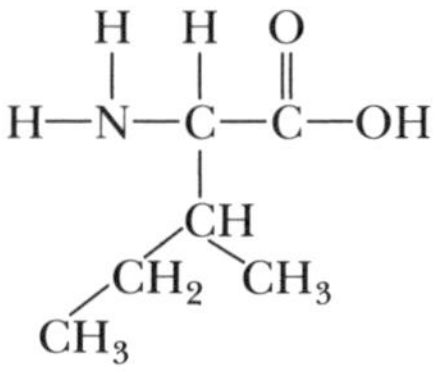

Serine

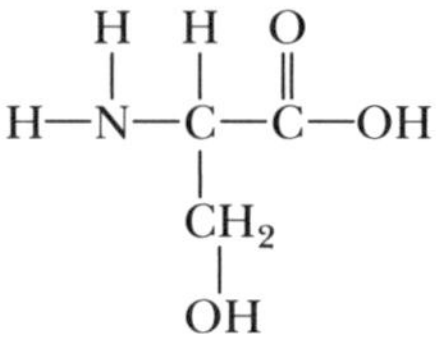

Threonine

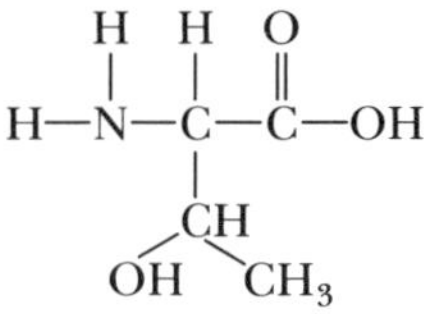

Aspartic acid

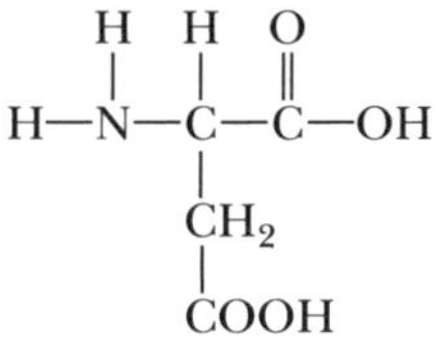

Glutamic acid

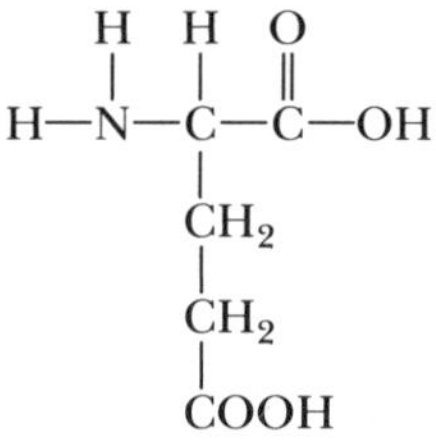

Lysine

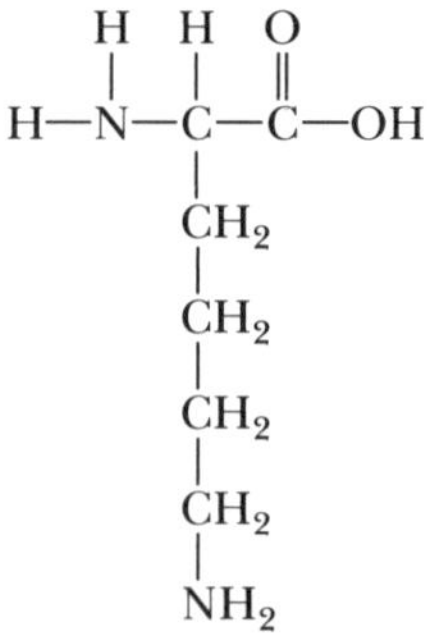

Arginine

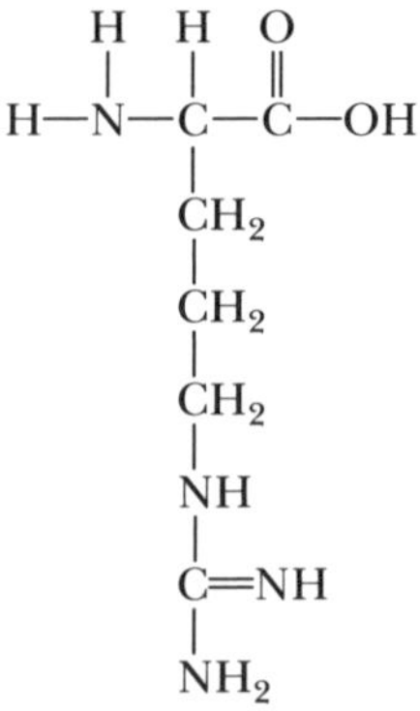

Asparagine

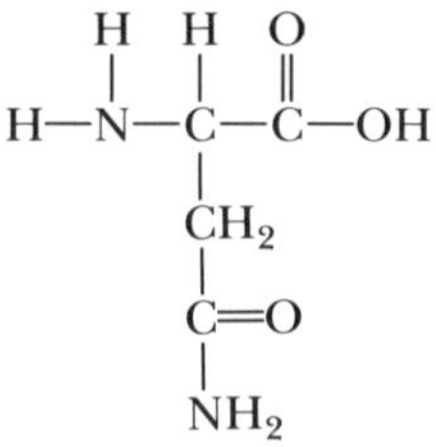

Glutamine

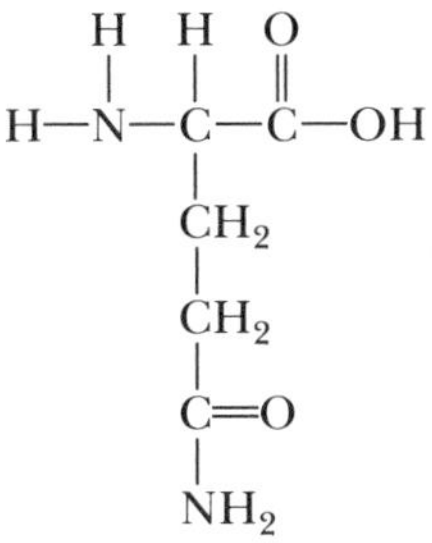

Cysteine

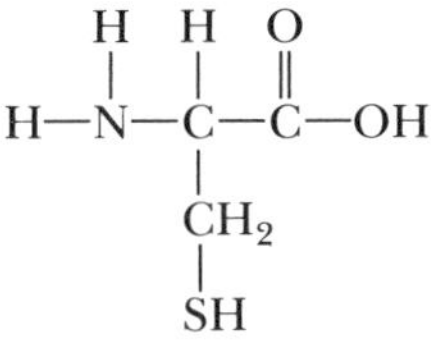

Methionine

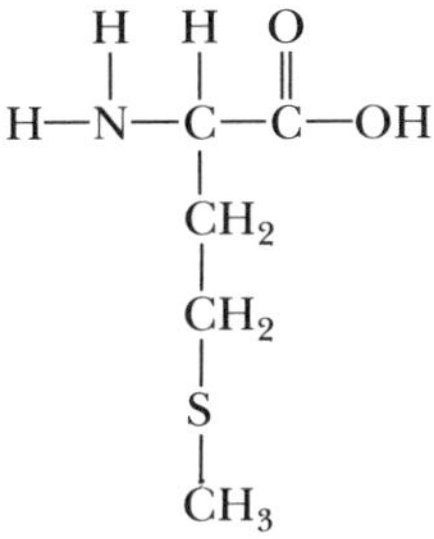

Phenylalanine

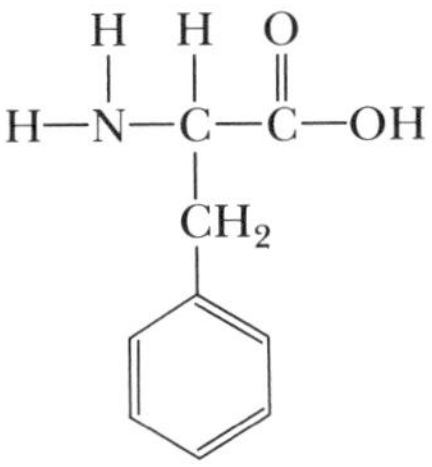

Tyrosine

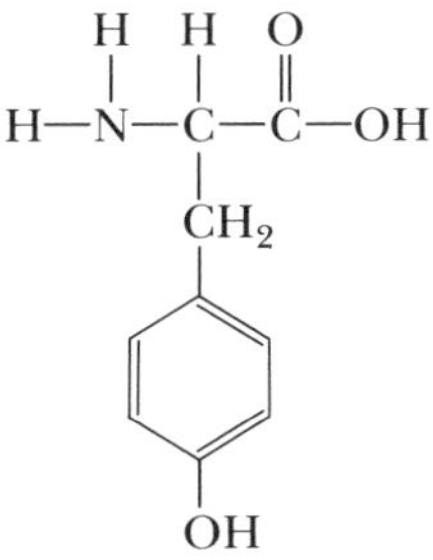

Tryptophan

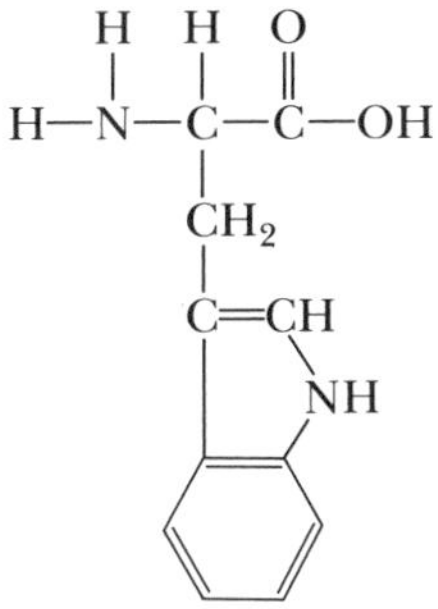

Histidine

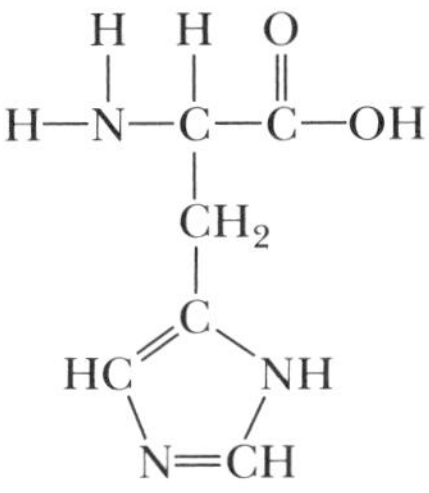

Proline

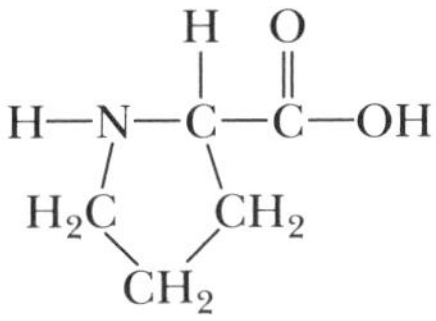

Figure 1-2 The 20 amino acids of proteins.

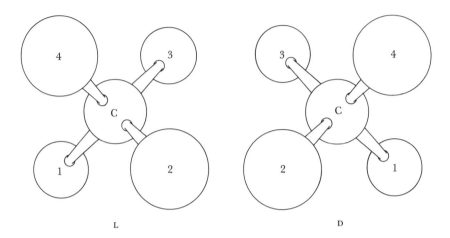

Figure 1-3 Optical isomerism results when four different radicals are attached to the same carbon atom. The two mirror images of the three-dimensional structure thus formed (dextro, D, and levo, L) are not superimposable, as this drawing shows. In the amino acids of proteins, the radicals attached to the central carbon atom are as follows: 1 = COOH; 2 = H; 3 = NH$_2$; 4 = any one of 20 different radicals. Protein amino acids have an L configuration, except for glycine, the simplest one, where 4 = H and the mirror images are indistinguishable. (This diagram is intended to show the spatial arrangement around the carbon atom, not the true relative sizes of atoms and radicals.)

acids occurred in the biochemistry of organisms, the chances of finding D or L amino acids would presumably be equal.

Typical proteins consist of one or more chains, called polypeptides, each containing several hundred amino acids linked end to end. All 20 amino acids are usually represented in each chain (see Figure 1-4). The chains fold themselves up into intricate three-dimensional arrangements, or conformations, often resembling a tangled thread. The special properties of protein molecules, both enzymes and nonenzymes, depend on this conformation. If the conformation is destroyed (a process referred to as denaturation), the protein loses its capacity to function, even though its amino acid chain(s) remain intact. Under the right conditions, denatured proteins may renature spontaneously, with restoration of their function. Such restoration shows that the three-dimensional con-

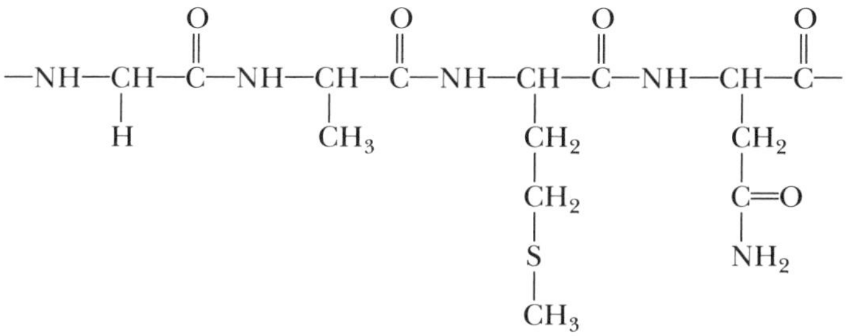

Figure 1-4 **Amino acids are joined together in chains to make proteins. Here a segment containing (from left to right) glycine, alanine, methionine, and asparagine is shown.**

figuration is determined by the amino acid sequence alone, and this sequence, we know, is encoded in the genes.

The rules by which nucleotides symbolize amino acid sequences are simple, but their demonstration was one of the great achievements of twentieth-century biology. In brief, the sequence of amino acids that characterizes any given polypeptide chain is represented in one particular gene, and this gene has no other function. A protein consisting of a single chain (or more than one chain, all identical in sequence) is encoded by a single gene; a protein consisting of two structurally different chains is encoded by two different genes; and so on. The manner of encoding goes like this. Each amino acid is represented by a set of three nucleotides of the four that make up DNA. Sixty-four sets of three can be made from four different nucleotides: AAA, AAG, AGA, and so forth, where the letters refer to the nucleic acid bases identified in Figure 1-1. Each such triplet represents one amino acid, except for three triplets that stand for message termination, or "nonsense." Sixty-one triplets are thus available for 20 amino acids, and most amino acids are consequently represented by two or more triplets in the genetic code, as shown in Table 1-1.

The genetic endowment of every organism therefore consists of a set of specifications, encoded in its DNA, for the synthesis of a large array of enzymes and other protein molecules. From this initial input, all other aspects of the organism follow—its development, structure, metabolism, and behavior—insofar as these are genetically determined. Thus nucleic acids and proteins form an

Table 1-1 The Genetic Code

Amino acid	Base triplets
Glycine	GGT, GGC, GGA, GGG
Alanine	GCT, GCC, GCA, GCG
Valine	GTT, GTC, GTA, GTG
Leucine	TTA, TTG, CTT, CTC, CTA, CTG
Isoleucine	ATT, ATC, ATA
Serine	TCT, TCC, TCA, TCG, AGT, AGC
Threonine	ACT, ACC, ACA, ACG
Aspartic acid	GAT, GAC
Glutamic acid	GAA, GAG
Lysine	AAA, AAG
Arginine	CGT, CGC, CGA, CGG, AGA, AGG
Asparagine	AAT, AAC
Glutamine	CAA, CAG
Cysteine	TGT, TGC
Methionine	ATG
Phenylalanine	TTT, TTC
Tyrosine	TAT, TAC
Tryptophan	TGG
Histidine	CAT, CAC
Proline	CCT, CCC, CCA, CCG
Nonsense	TAA, TAG, TGA

interlocking and interdependent system: the production of both kinds of molecules depends on the activities of large numbers of enzymes, and the synthesis of enzymes requires the information contained in DNA. Whatever is unique about living matter is inherent in this self-sustaining system, the "genetic system."

The relation between genes and proteins is somewhat convoluted but understandable. The organism must synthesize a great variety of different kinds of proteins in order to survive. But protein molecules are large, highly improbable structures that must be built up from individual amino acids. If every organism had to discover for itself how to assemble amino acids in the correct order to produce useful proteins, survival would be impossible. This information—the essential, irreplaceable genetic heritage—must be transmitted from parent to offspring. If amino acid sequences could be copied from preexisting protein molecules, nucleic acids would not be needed. But proteins are structurally unsuited for copying. Nucleotide sequences, on the other hand, can readily be copied from polynucleotides. Hence specifications for the assembly of protein molecules are encoded in nucleic acids, and it is nucleic acids that are duplicated in each generation and transmitted by inheritance.

Proteins and nucleic acids by themselves do not constitute an organism, of course. Enzymes need raw materials to work on in order to synthesize more nucleic acids, more enzymes, and all the other substances that go to make an organism. They also need an energy source and a solvent. The solvent, water, is actually the major constituent of most living things. (More about energy sources and water appears later in this book.) The point is that given raw materials, energy, and water, the genetic system has the capacity to produce an organism, including all the structures that have no genetic properties themselves, such as the membrane that encloses each cell.

Beyond these basic requirements, to form an organism the system needs a program that determines the order of events. The thousands of genes that contain the program for a living system are not all active at the same time. Especially in multicellular animals and plants, a complex, stereotyped series of changes called development takes place, in the course of which different genes are called into play at different times in different cells. To take a simple example, only certain cells of the body produce hemoglobin, and the genes that carry the information needed for constructing the two kinds of amino acid chains that make up this protein, although present in all cells, are activated only in hemoglobin-producing cells. Furthermore, the mammalian fetus makes a different hemoglobin than the adult does. This means that different hemoglobin genes are brought into play at different stages of development. Decisions of this kind, extended over all the genes and all the cells of the organism, bring about the formation of the individual animal or plant starting at fertilization of the egg. The program of controls is encoded in the genes. The nature of the control signals and the various mechanisms involved in development are not yet well understood and are the subject of much current research.

What is the source of the information contained in the genes? The immediate source is the genes of the parents. The ultimate source of genetic information, however, is chance mutations—random changes of individual nucleotides or sometimes larger rearrangements of DNA—screened by natural selection. Mutant genes replicate like any other genes, but when translated they give rise to proteins with new amino acid sequences and new properties, or to modified developmental programs. In most cases, mutations are valueless or harmful, and they are discarded by natural selection. Occasionally, however, a mutation results in a useful new protein or developmental maneuver, one that is advantageous to the individuals carrying it. Such a mutation is preserved and spread by natural selection—that is, individuals who carry the mutation leave more offspring, on the average, than individuals who do not. Eventually, the mutant type may come to dominate the population.

The evolution of novel proteins accounts for the recent acquisition of resistance to insecticides in insects, mites, and ticks, a phenomenon observed worldwide. In some insect species, a mutant form of acetylcholinesterase, an essential enzyme of nerve cells, is insensitive to the organic phosphates designed to poison it. A recently evolved enzyme, dehydrochlorinase, destroys DDT, thereby conferring resistance to the chemical in houseflies and mosquitoes. Novel proteins occur quite commonly, not only in insects but also in bacteria, where evolution of antibiotic resistance occurs so frequently that it constitutes a public-health problem. Studies have shown that insecticides and antibiotics do not cause mutations conferring resistance. Such mutations already exist at a low frequency in unexposed populations. Once toxic compounds have killed off sensitive individuals, the resistant few expand in numbers and replace the original type.

These are examples of small evolutionary steps that have occurred in recent times. The totality of the genetic information represented by any species is the product of a very long history of such steps. Thus the genes contain a record of mutational discoveries that extends to the remotest past.

Now we can answer the question "What is life?" The genetic attributes of living things—that is, the capacities for self-replication and mutation—underlie the evolution of all the structures and functions that distinguish living objects from inanimate ones. Therefore, we can answer the question like this: Life is synonymous with the possession of genetic properties. Any system with the capacity to mutate freely and to reproduce its mutations must almost inevitably evolve in directions that will ensure its preservation. Given sufficient time, the system will acquire the complexity, variety, and purposefulness that we recognize as "alive." The designer whose hand is seen everywhere in the living world thus turns out to be the cumulative effects of natural selection acting on spontaneous mutations over long ages of time.

The genetic view of the nature of life was first stated clearly by an American, H. J. Muller (1890–1967), the discoverer of the mutagenic effect of X-rays and one of the founders of modern genetics. Muller presented his paper, "The Gene as the Basis of Life," to an international congress in 1926, long before anything was known about the chemical nature of the genes or their relation to proteins. Muller perceived, however, that the properties of self-replication and mutability were at the center of the phenomenon called life. His closely reasoned argument does not lend itself to brief quotation, but this excerpt from the résumé of the paper suggests Muller's logic and style:

> It is pointed out that growth involves a specific autocatalysis, without which life cannot exist. The gene, when in its protoplasmic setting, is known to possess this property of "specific autocatalysis." Still more remarkable, the gene can mutate without losing its specific autocatalytic power. In view of this latter peculiarity of the gene, it becomes a supererogation, and involves improbable assumptions, to suppose that any other portion of the protoplasm, independently of the gene, is specifically autocatalytic; "growth" of the rest of the protoplasm would follow as a by-product of gene activity. Likewise it involves apparently insuperable difficulties to suppose that, in the most primitive living matter, highly organized companion substances

> to the gene ("protoplasmic" in nature) were necessary
> to make the gene-autocatalysis possible. Genes (simple
> in structure) would, according to this line of reason-
> ing, have formed the foundation of the first living
> matter. By virtue of their property (found only in "liv-
> ing" things) of mutating without losing their growth
> power they have evolved even into more complicated
> forms, with such by-products—protoplasm, soma,
> etc.—as furthered their continuance. Thus they
> would form the basis of life.

The genetic view is now widely, if not universally, accepted. Nongenetic definitions of life tend to be vague or too inclusive. Crystals and flames, for example, are hard to exclude from such definitions: crystals show the properties of growth and order in a high degree, and they can reproduce themselves by seed crystals; flames not only grow and reproduce (by means of sparks), they also maintain themselves through an active metabolism.

The genetic definition of life allows us to draw a conclusion of far-reaching importance. Since the genes and proteins of all species are constructed out of the same nucleotides and amino acids, and since the genetic code is, with minor exceptions, also universal, all terrestrial organisms are fundamentally alike. Despite appearances, there is only one form of life on the earth, and this life need have originated only once.

LIFE AND CARBON CHEMISTRY

The genetic-evolutionary view carries implications concerning the chemical composition of living matter that we must now examine. Aside from the inherent interest of the subject, the possible chemistry of extraterrestrial organisms is clearly important for anyone planning a search for life on Mars. Life on our planet is based on the chemistry of carbon. Compounds of carbon with a few other elements, principally the light elements hydrogen, nitrogen, and oxygen, form the substances of the genetic system (see Table 1-2). The question we must now ask is whether any other chemical element could replace carbon to generate a biochemistry. Although science-fiction writers frequently tell us yes, it is by no means clear that any other element could substitute for carbon.

Table 1-2   Elemental Composition of Proteins and DNA

	Number of atoms per 100 atoms	
Element	314 proteins (average)*	Human DNA
Carbon (C)	31.6	29.8
Hydrogen (H)†	49.6	37.5
Oxygen (O)	9.7	18.3
Nitrogen (N)	8.8	11.3
Sulfur (S)	0.3	–
Phosphorus (P)	–	3.1

* Recalculated from Dayhoff (1972).

† There are more hydrogen atoms than any others in these molecules, but hydrogen does not form the framework of the structures, because it can make only one covalent bond.

The properties of the carbon atom that peculiarly fit it for the construction of seemingly limitless numbers of large, complex, yet stable molecules have often been noted by chemists. First, carbon has a unique ability to make four strong chemical bonds, called covalent bonds, with other atoms, including other carbon atoms. Because covalent bonds have direction in space, carbon atoms can form the skeletons of immense three-dimensional structures with definite architectures, such as proteins and nucleic acids.

Next is the inertness of the compounds of carbon. Organic compounds are thermodynamically unstable under the conditions that prevail on the earth's surface. They are not at equilibrium with the environment, but like a stone lodged on the side of a mountain, if given a strong enough push they will attain equilibrium by rolling downhill, so to speak. Thus, if activated by a catalyst or by heat, organic matter combines with the oxygen of the atmosphere; many organic compounds react with water as well, or undergo a variety of other changes. Even though they are thermodynamically unstable, however, carbon compounds are characteristically inert—that is, slow to react. Tetravalent carbon atoms do not easily find reaction pathways, and this impedes the attainment of thermodynamic equilibrium—in the terms of our analogy, the stone on the mountainside is in a deep hole. This inertness, which stems from the electronic structure of the carbon atom, explains why carbon atoms can form highly elaborate yet long-lived molecular structures. When the biological scheme of things requires action, enzymes combine with the molecules to be altered and provide the reaction pathways needed.

These properties fit carbon uniquely for the construction of genetic systems. They explain why carbon forms more compounds than all other elements combined. They also explain why carbon, which composes only 0.5 percent of the earth's crust, is the characteristic element of living matter rather than silicon, for example, a close chemical relative. For every atom of carbon at the earth's surface there are 25 atoms of silicon, yet silicon plays a very minor role in biochemistry. Like carbon, silicon forms four covalent bonds, but unlike carbon, the strength of these bonds varies. The silicon-silicon bond is weak and the silicon-oxygen bond is strong. Therefore most of the silicon on the earth occurs in silicates—large, inert compounds in which each silicon atom is bonded to four oxygen atoms—and chains of silicon atoms containing more than six or so atoms are unknown, in marked contrast to the large structures found in carbon chemistry. Compounds of silicon and hydrogen, called silanes, are also quite different from their carbon-containing homologs, the hydrocarbons. Whereas hydrocarbons are inert, silanes ignite on contact with air, and they decompose in water. So reactive are they, in fact, that the first prerequisite for a chemist aspiring to synthesize higher silanes is said to be a great deal of courage. Again, the properties of silanes are understandable in the light of the electronic structure of the silicon atom; and they explain why silicon is important in rocks but not in the living world.

Carbon is so superior for the building of complex molecules that the possibility of forming genetic systems with other elements has never seriously been considered. It has been pointed out that chains of other elements (for example, of alternating silicon and oxygen atoms, $-Si-O-Si-O-$) are potentially capable of storing information, but information storage is only one of the functions that a living system must perform. The others are mutation, replication, and utilization of that information. While we cannot prove that no element but carbon can accomplish these functions, the possibility appears so remote that we shall assume carbon to be unique in this respect. This does not imply, of course, that the genetic systems of extraterrestrial species must be chemically identical with our own, only that they must be built out of carbon compounds. This conclusion has far-reaching consequences for the possibility of life on other planets, as we shall see.

Some may find it disappointing and perhaps even a little depressing to conclude that the surest way to find life on another world is to search for complex chemical systems based on carbon. After all, that is what we have on earth. Is there no hope for exotic creatures formed from, say, vanadium, molybdenum, or praseodymium? I think not. Such elements are not only chemically unsuitable, they are rare, whereas carbon is one of the most abundant elements in the universe. To the extent that chance enters into the origin of life—and that is the subject of the next two chapters— abundant elements are more likely to be involved than rare ones, other things being equal. In this case, other things are not equal. The great versatility of the carbon atom makes it the element most likely to provide solutions—even exotic solutions—to the problems of survival on other planets.

CHAPTER TWO

The Origin of Life:
Spontaneous Generation
and Panspermia

It's hard to make a good theory; a theory has to be reasonable, but a fact doesn't.

George W. Beadle, Geneticist,
Nobel Prize in Physiology or Medicine, 1958

The physicist Philip Morrison once remarked that the discovery of life on another planet would transform life from a miracle into a statistic. It would certainly add something to our knowledge of the origin of life. It could help us to answer a variety of questions that cannot be approached in any other way. It would test the belief that life must be based on carbon chemistry, and if the newfound life were carbon-based as expected, we might discover whether genetic systems can be constructed out of molecules other than nucleic acids and proteins that are built up from their familiar subunits. We might also answer the perennial question of whether or not any other solvent can substitute for water in a living system. And so on through a long list of riddles.

If the new organisms were different from us chemically in some fundamental way, we would know that life had arisen independently at least twice in the solar system. But if they were like us, with similar proteins and nucleic acids, with the same optical isomerism, and with the same genetic code, we would have a new problem. Should we conclude that two origins of life had occurred, or that life originated once and was transported from one planet to the other? The latter seems much more likely. Whatever the actual findings, the discovery of extraterrestrial life would obviously be of enormous interest from the viewpoint of fundamental biology.

Only three naturalistic theories about the origin of life have held sway since the time of Aristotle. (Naturalistic ideas differ from supernaturalistic ones in being testable and therefore within the realm of science.) The theories are spontaneous generation, panspermia, and chemical evolution. These form an important part of the historical and intellectual background of the search for life in the solar system. Chemical evolution, the modern theory and the subject of the next chapter, is still under investigation.

SPONTANEOUS GENERATION

Spontaneous generation is the name given to the notion that living things continually arise spontaneously from nonliving material—mud, dew, or decaying organic matter, for example. It also includes cases in which one form of life was claimed to be directly transformed into a different form—grain turning into mice, for example. From the time of Aristotle (384–322 B.C.) to the middle of the seventeenth century, the spontaneous generation of plants and animals was commonly accepted as fact. For the next two centuries, microbes replaced higher forms of life as the supposed products of spontaneous generation.

The literature of these ages contains many recipes for producing worms, mice, scorpions, eels, and, later, microorganisms de novo. In many cases, these descriptions cite the authority of ancient Greek and Arabic writers; less frequently, circumstantial reports detail actual experiments.

Historians tell us that the Greeks created science and that Aristotle was the father of biology. Indeed, Aristotle did bring the rational Greek spirit, the conviction that human beings can comprehend nature by the application of reason, to the study of living things. In his philosophical writings, Aristotle was much concerned with methods of logical proof: he founded formal logic and, among other things, invented the syllogism. He also observed nature, especially living nature, but in this role he was unreliable. Although some of his descriptions, such as those dealing with the behavior of animals, have been much admired, his biological writings in general contain numerous errors in fact, as well as some large misconceptions. Much of what he reports was probably hearsay.

Aristotle discusses spontaneous generation in the passage from *Historia Animalium* that follows:

> Now there is one property that animals are found to have in common with plants. For some plants are generated from the seed of plants, while other plants are self-generated through the formation of some elemental principle similar to a seed; and of these latter plants some derive their nutriment from the ground, while others grow inside other plants, as is mentioned, by the way, in my treatise on botany. So with animals, some spring from parent animals according to their kind, while others grow spontaneously and not from kindred stock; and of these instances of spontaneous generation some come from putrefying earth or vegetable matter, as is the case with a number of insects, while others are spontaneously generated in the inside of animals out of the secretions of their several organs. . . . But whensoever creatures are spontaneously generated, either in other animals, in the soil, or on plants, or in the parts of these, and when such are generated male and female, then from the copulation of such spontaneously generated males and females there is generated a something—a something never identical in shape with the parents, but a something imperfect. For instance, the issue of copulation in lice is nits; in flies, grubs; in fleas, grubs egglike in shape, and from these issues the parent species is never reproduced, nor is any animal produced at all, but the like nondescripts only.

Aristotle well knew that many insects have a complex life cycle that takes them through a larval and pupal stage before they become adults. But while he seems oblivious to the factual errors in his story of two kinds of insect genesis, he is acutely sensitive to its logical requirements. Spontaneous generation would not make sense—indeed, its existence would be questionable—if the creatures so formed could reproduce normally. Hence, says Aristotle, these creatures produce only sexless "nondescripts" when they mate, thus ensuring a continual need for spontaneous generation.

This is all nonsense, of course, but the science of Aristotle was science in its infancy. It is enough that Aristotle considered the study of insects a worthy occupation. What is hard to believe is that Aristotle's views survived essentially unchanged for nearly 2000 years. Even the medieval Church accepted Aristotle's authority on spontaneous generation, and St. Thomas Aquinas (1225–1274) himself reconciled Aristotle's teaching with Christian doctrine by declaring that spontaneous generation was brought about by angels acting through the agency of the sun.

An age of many superstitions, the sixteenth century marks the high point of the classical doctrine of spontaneous generation. The physician Paracelsus (1493–1541) strongly advocated it, and his follower, Jean Baptiste van Helmont (1577–1644), was the author of a much quoted method for producing mice from grains of wheat placed in a jar with an old undershirt. As Pasteur said, commenting two centuries later on van Helmont's earnest description: "What this proves is that it is easy to do experiments, but hard to do them flawlessly."

In a work entitled *Natural Magick,* first published in 1558, Giambattista della Porta recounts yet more sixteenth-century notions about spontaneous generation. The author, a Neapolitan amateur scientist, was a founder and vice president of the Accademia dei Lincei, one of the earliest scientific societies. His popular book, a collection of technical arts, natural wonders, and practical jokes, was translated into several languages. These excerpts are from the English edition published in London in 1658.

> In Dariene, a Province of the new world, the air is most unwholesome, the place being muddy and full of stinking marishes; nay, the village is itself a marish, where Toads are presently gendred of the drops

wherewith they water their houses, as *Peter Martyr* writes. A Toad is likewise generated of a duck that hath lyen rotting under the mud, as the verse shews which is ascribed to the duck; When I am rotten in the earth, I bring forth Toads. . . .

Florentinus the Grecian saith, That Basil chewed and laid in the Sun, will engender serpents. *Pliny* addeth; that if you rub it, and cover it with a stone, it will become a Scorpion; and if you chew it, and lay it in the Sun, it will bring forth worms. . . .

Salamander is gendred of the water; for the Salamander itself genders nothing, neither is there any male or female amongst them, nor yet amongst Eeeles. . . .

So the fish called *Ortica,* and the Purple, and Muscles, and Scallops, and Perwinkles, and Limpins, and all Shel-fish are generated of mud; for they cannot couple together, but live only as plants live. And look how the mud differs, so doth it bring forth different kinds of fishes: durty mud genders Oysters, sandy mud Perwinkles, the mud of the Rocks breedeth Holoturia, Lepades, and such-like. Limpins, as experience hath shewed, have bred of rotten hedges made to fish by; and as soon as the hedges were gone, there have been found no more Limpins.

To a modern reader, accustomed to think of the origin of life as a singular event, the most remarkable in the history of the earth, these writings have a fairy-tale quality. Yet there is no reason to suppose that anyone made up such stories. It is much more likely that these confident accounts—to the extent that they were based on actual observation—arose from misinterpretations of commonplace events, readily accepted because they harmonized with the writings of the ancients and with the common wisdom.

Along with many another time-honored illusion, the classical doctrine of spontaneous generation died in the Renaissance. The man who overthrew it was Francesco Redi (1626–1697), by all accounts an ideal late-Renaissance figure: a practicing physician, a popular poet, and one of the first modern biologists. Redi's healthy scepticism, keen powers of observation, and delightful writing style are visible in *Experiments on the Generation of Insects* (1668), the book on which his scientific reputation chiefly rests. Although insects were his main subject, he also investigated the generation of scorpions, toads, spiders, frogs, and quail. Not only was he unable to confirm the popular belief that these creatures could be generated spontaneously, in most cases he demonstrated their true origin in fertilized eggs. Thus his careful experiments swept away the accumulated lore of 20 centuries.

Redi cast his work in the form of a letter to his friend, Carlo Dati. After a historical introduction, he continues:

> It being thus, as I have said, the dictum of ancients and moderns, and the popular belief, that the putrescence of a dead body, or the filth of any sort of decayed matter engenders worms; and being desirous of tracing the truth in the case, I made the following experiment:
>
> At the beginning of June I ordered to be killed three snakes, the kind called eels of Aesculapius. As soon as they were dead, I placed them in an open box to decay. Not long afterwards I saw that they were covered with worms of a conical shape and apparently without legs. These worms were intent on devouring the meat, increasing meanwhile in size, and from day to day I observed that they likewise increased in number. . . .

There follows a minute description of the "worms" and their metamorphosis into pupae and finally adult flies. Redi carefully describes the results of his repeated observations of different kinds of meat and continues:

> I continued similar experiments with the raw and cooked flesh of the ox, the deer, the buffalo, the lion, the tiger, the dog, the lamb, the kid, the rabbit; and sometimes with the flesh of ducks, geese, hens, swallows, etc., and finally I experimented with different kinds of fish. . . . In every case, one or other of the above-mentioned kinds of flies were hatched, and

sometimes all were found in a single animal . . . and almost always I saw that the decaying flesh and the fissures in the boxes where it lay were covered not alone with worms, but with the eggs from which, as I have said, the worms were hatched. These eggs made me think of those deposits dropped by flies on meats, that eventually become worms, a fact noted by the compilers of the dictionary of our Academy, and also well known to hunters and to butchers, who protect their meats in Summer from filth by covering them with white cloths. . . .

Having considered these things, I began to believe that all worms found in meat were derived directly from the droppings of flies, and not from the putrefaction of the meat, and I was still more confirmed in this belief by having observed that, before the meat grew wormy, flies had hovered over it, of the same kind as those that later bred in it. Belief would be vain without the confirmation of experiment, hence in the middle of July I put a snake, some fish, some eels of the Arno, and a slice of milk-fed veal in four large, wide-mouthed flasks; having well closed and sealed them, I then filled the same number of flasks in the same way, only leaving these open. It was not long before the meat and the fish in these second vessels became wormy, and flies were seen entering and leaving at will; but in the closed flasks I did not see a worm, though many days had passed. . . .

Leaving this long digression and returning to my argument, it is necessary to tell you that although I thought I had proved that the flesh of dead animals could not engender worms unless the semina of live ones were deposited therein, still to remove all doubt, as the trial had been made with closed vessels into which the air could not penetrate or circulate, I wished to attempt a new experiment by putting meat and fish in a large vase closed only with a fine Naples veil that allowed the air to enter. For further protection against flies, I placed the vessel in a frame covered with the same net. I never saw any worms in the meat, though many were to be seen moving about on the net-cov-

> ered frame. These, attracted by the odor of the meat,
> succeeded at last in penetrating the fine meshes and
> would have entered the vase had I not speedily re-
> moved them.

The tenor of this is very modern. The last two experiments have
become classics that served as models for all subsequent investigations of spontaneous generation. In the rest of *Experiments on the
Generation of Insects,* Redi describes more experiments and provides a running critique of popular myths and confusions relating
to the generation of animals. Della Porta comes in for comment
halfway through the work:

> This was a favorable opportunity for proving the state-
> ment of Batista Porta, that the toad is generated from
> a duck putrefying on a dung-heap. Three experiments
> with this material brought no result, hence I was con-
> vinced that Porta, otherwise a most interesting and
> profound writer, had been too credulous.

As Redi's book went through five editions in 20 years, belief in
the spontaneous generation of animals gradually died out among
educated people. The question was reopened at another level,
however, following the discovery of microorganisms around 1675
by Anton Leeuwenhoek (1632–1723), a Dutchman. This achieve-
ment was made possible by improvements in the art of lens making
in the seventeenth century. Leeuwenhoek was himself a skilled lens
maker, as well as an avid microscopist. Indeed, based on the
number of important discoveries he made in the course of his long
life, Leeuwenhoek can be considered the greatest microscopist of
all time.

Microbes are so small and seemed so simple that from the time of
their discovery they were widely believed to be the products of
decay, existing in a twilight zone between life and nonlife. Thus renewed, the debate about spontaneous generation erupted in a
famous eighteenth-century controversy between an English cler-
gyman, J. T. Needham (1713–1781), and an Italian physiologist,
Abbe Lazaro Spallanzani (1729–1799). Needham claimed that

mutton gravy and similar infusions, first heated and then sealed in a vessel with a small amount of air, would, in the course of a few days, invariably produce microorganisms and decompose. The heat, he believed, had killed all preexisting organisms, and therefore this result demonstrated spontaneous generation. Repeating Needham's experiments, Spallanzani showed that if the flasks were heated after sealing, nothing grew in them, nor did the matter in them putrefy, no matter how long it was kept. (In one of his experiments, Spallanzani sealed green peas with some water in a glass vessel that was then held in boiling water for 45 minutes. This procedure was used in 1804 by François Appert, a Parisian chef, to make the first preserved foods. The canning industry is thus one of the by-products of the spontaneous generation controversy.)

Needham replied that excessive heating had destroyed a vital element in the air within the sealed vessels, without which spontaneous generation could not occur. The science of gas analysis was not sufficiently advanced to settle this point. Indeed, the issue raised by Needham turned out to be a tricky one that was not cleared up for a century. Some of the most illustrious figures of nineteenth-century science, including Joseph Louis Gay-Lussac, Theodor Schwann, Hermann von Helmholtz, Louis Pasteur, and John Tyndall, became involved in the debate. Gay-Lussac, the great French chemist, gave support to Needham by showing that oxygen disappears from air heated in the presence of organic matter, and further experiments seemed to show that absence of oxygen is the necessary condition for preservation of foods. But the crucial experiment—that is, the Redi experiment but applied to microorganisms—remained beyond reach.

The question appeared to be simple: Will a sterilized organic infusion remain free of microbial growth in the presence of air from which all microbes have been removed but which is otherwise unaltered? Simple though the question seemed, a decisive answer required a technology that had not yet been invented. Many ingenious experiments were performed, but investigators obtained only incorrect, or partially correct, solutions, and disagreed widely among themselves. Because of the philosophical and practical importance of the question of spontaneous generation, feelings ran deep, and the debate generated considerable passion.

The matter came to a climax in 1859, when Felix Pouchet (1800–1872), director of the Museum of Natural History at Rouen, published a book in which experimental proof of spontaneous generation was once again claimed. Pouchet opened his preface with the ominous words: "When, by meditation, it became evident to me that spontaneous generation was still one of the means that nature employs for the reproduction of beings, I applied myself to discover by what means one could succeed in demonstrating the relevant phenomena." John Tyndall (1820–1893), the British physicist who was to make an important contribution to the debate, commented on Pouchet's entry into the arena as follows:

> Never did a subject require the exercise of the cold critical faculty more than this one—calm study in the unravelling of complex phenomena, care in the preparation of experiments, care in their execution, skilful variation of conditions, and incessant questioning of results until repetition had placed them beyond doubt or question. To a man of Pouchet's temperament the subject was full of danger—danger not lessened by the theoretic bias with which he approached it.

At this point Louis Pasteur (1822–1895) became involved. In studying alcoholic fermentation, Pasteur, a chemist, had concluded, against much opposition, that fermentations were caused by living organisms. This experience was good preparation for the problem he now undertook. In essence, Pasteur's investigation, a technically brilliant series of experiments and one of the monuments of biology, essentially brought to a close the long history of the spontaneous generation controversy. In it, Pasteur resolved all of the difficulties that had plagued earlier workers. He demonstrated unequivocally that the mysterious principle in the air that causes microbial life to appear in sterilized broths is microbial life itself, carried on dust particles.

Let us briefly examine just one of Pasteur's experiments, the simplest and most elegant one, the effectiveness of which surprised even Pasteur. He placed a suitable medium—yeast extract with added sugar, for example—into a flask and then pulled the neck of

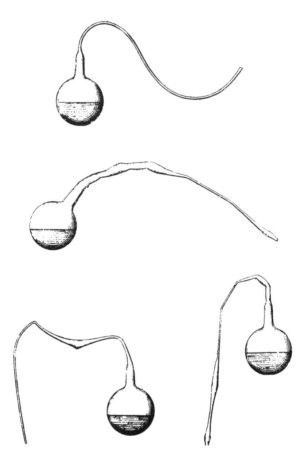

Figure 2-1   Pasteur's "swan-necked" flasks, from his 1862
memoir on spontaneous generation.

the flask out over a flame to form a narrow—but open—tube,
which he bent in various ways, as shown in Figure 2-1. He next
brought the medium to a boil for a few minutes and then let the
flask cool down. The medium in such flasks remained sterile indefinitely, even though it was in contact with the air. To Pasteur's surprise, the flasks could even be moved from place to place without
becoming infected. He explained this by pointing out that the air in
the long neck acts as a cushion preventing rapid air movements, so
that entering dust falls out and is caught on the walls of the neck

before it gets to the medium. To show that the solution would actually support the growth of microorganisms if inoculated, Pasteur cut off the necks of some of the flasks, and they soon developed microbial growths.

Pasteur thus succeeded in repeating the Redi experiment at the level of single-celled organisms. He showed that in the microbial world, as among higher forms, all life comes from preexisting life. Pasteur's study was definitive, but various rumblings and counterattacks continued for some years. A particularly interesting episode took place in England in the 1870s when John Tyndall defended Pasteur's position against an attack by a physician named H. Charlton Bastian. Tyndall's studies of the scattering of light by particles in the atmosphere enabled him to design novel experiments demonstrating the role of dust in the transport of infection. He showed that putrescible media in open test tubes would remain sterile as long as the air above them was dust-free. The following description of how he detected the presence of dust in the air explains Tyndall's reputation as the outstanding popularizer of science of the Victorian era:

> Let us now return to London and fix our attention on the dust of *its* air. Suppose a room in which the housemaid has just finished her work to be completely closed, with the exception of an aperture in a shutter through which a sunbeam enters and crosses the room. The floating dust reveals the track of the light. Let a lens be placed in the aperture to condense the beam. Its parallel rays are now converged to a cone, at the apex of which the dust is raised to almost unbroken whiteness by the intensity of its illumination. Defended from all glare, the eye is peculiarly sensitive to this scattered light. The floating dust of London rooms is organic, and may be burned without leaving visible residue. The action of a spirit-lamp flame upon the floating matter has been elsewhere thus described:[1]

---

[1] By Tyndall himself.—*Author's note.*

"In a cylindrical beam which strongly illuminated the dust of our laboratory, I placed an ignited spirit-lamp. Mingling with the flame, and round its rim, were seen curious wreaths of darkness resembling an intensely black smoke. On placing the flame at some distance below the beam, the same dark masses stormed upwards. They were blacker than the blackest smoke ever seen issuing from the funnel of a steamer; and their resemblance to smoke was so perfect as to prompt the conclusion that the apparently pure flame of the alcohol-lamp required but a beam of sufficient intensity to reveal its clouds of liberated carbon.

But is the blackness smoke? This question presented itself in a moment, and was thus answered: A red-hot poker was placed underneath the beam; from it the black wreaths also ascended. A large hydrogen flame, which emits no smoke, was next employed, and it also produced with augmented copiousness those whirling masses of darkness. Smoke being out of the question, what is the blackness? It is simply that of stellar space; that is to say, blackness resulting from the absence from the track of the beam of all matter competent to scatter its light. When the flame was placed below the beam, the floating matter was destroyed *in situ;* and the heated air, freed from this matter, rose into the beam, jostled aside the illuminated particles, and substituted for their light the darkness due to its own perfect transparency. Nothing could more forcibly illustrate the invisibility of the agent which renders all things visible. The beam crossed, unseen, the black chasm formed by the transparent air, while, at both sides of the gap, the thick-strewn particles shone out like a luminous solid under the powerful illumination."

Tyndall also invented a method, still known as tyndallization, for sterilizing solutions containing bacterial spores that can survive boiling water. In this procedure, a solution is heated on successive days: ungerminated spores survive the heat but germinated spores do not, and the solution becomes sterile after a few heatings. Altogether, Tyndall's experiments were so original and his espousal of

Pasteur's cause so vigorous that he shares with Pasteur the credit for overthrowing the doctrine of spontaneous generation.

Another practical benefit now emerged from the controversy. It occurred to Joseph Lister (1827–1912), a surgeon and a contemporary of Pasteur and Tyndall who knew of their work, that if he could keep the germs in the air of his surgery from reaching his patients' incisions, he might be able to lower the incidence of infection. At the time, between 25 and 50 percent of patients died following amputations in British hospitals, chiefly from infection. Things were much worse on the battlefield: of 13,000 amputations performed by French surgeons during the Franco-Prussian war, no less than 10,000 are said to have been fatal! As long as germs were believed to be generated spontaneously, there was no reason to exclude them from wounds. On the basis of Pasteur's findings, however, Lister decided to try to kill bacteria before they reached the area of the incision, and he hit on the use of carbolic acid (phenol) as an antibacterial agent. He sterilized his instruments, sprayed the surgery, and saturated the patient's dressings with a phenolic solution. These measures were strikingly successful, and antiseptic surgery was born.

PANSPERMIA

The doctrine of spontaneous generation had been collapsing for centuries, and its burial by Pasteur and Tyndall can hardly have come as a surprise to contemporary scientists. Nevertheless, there was no other theory to take its place. One can imagine that in the nineteenth century, with so little known about the chemical organization of living matter, anyone who tried to think about the origin of life had to conclude that the effort was futile. As Darwin remarked in a letter to Hooker in 1863, "It is mere rubbish, thinking at present of the origin of life; one might as well think of the origin of matter."

Darwin was right. Not enough was known about the nature of life or the history of the earth to speculate about the origin of life profitably. To some eminent scientists, however, the demise of spontaneous generation suggested that life had no origin but, like matter and energy, had always existed. According to this view, germs of life drifted from place to place in cosmic space: when they fell on a favorable planet, they initiated biological evolution. Hermann von Helmholtz (1821–1894) and William Thomson, later Lord Kelvin (1824–1907), two of the most celebrated physicists of the nineteenth century, espoused this idea. Helmholtz, who had himself experimented on the spontaneous generation of bacteria, said in a lecture delivered in 1871:

> I cannot object if anyone considers this hypothesis to be in a high, or even in the highest, degree improbable. But to me it seems a perfectly correct scientific procedure, that when all our attempts fail in producing organisms from inanimate matter, we may inquire whether life has ever originated at all or not, and whether its germs have not been transported from one world to another, and have developed themselves wherever they found a favourable soil.

Helmholtz and Thomson were close friends, and they may well have discussed the matter. In any case, Thomson made a very similar proposal in his presidential address to the British Association for the Advancement of Science a few months later:

> Careful enough scrutiny has in every case up to the present day discovered life as antecedent to life. Dead matter cannot become living without coming under the influence of matter previously alive. This seems to me as sure a teaching of science as the law of gravitation. . . . I am ready to adopt, as an article of scientific faith, true through all space and through all time, that life proceeds from life, and from nothing but life. How, then, did life originate on the Earth?

He then argued that many other life-bearing worlds must exist in the universe, that these must occasionally break up following a collision with other bodies, with the result that fragments carrying living plants and animals are scattered through space.

> Hence . . . we must regard it as probable in the highest degree that there are countless seed-bearing meteoric

> stones moving about through space. If at the present
> instant no life existed upon this Earth, one such stone
> falling upon it might, by what we blindly call *natural*
> causes, lead to its becoming covered with vegeta-
> tion. . . . The hypothesis that life originated on this
> Earth through moss-grown fragments from the ruins
> of another world may seem wild and visionary; all I
> maintain is that it is not unscientific.

This idea was elaborated in 1908 by the Swedish chemist Svante
Arrhenius (1859–1927). He called the theory panspermia. To the
suggestions advanced by Helmholtz and Kelvin, he added a few of
his own. Bacterial spores and virus particles, he proposed, could be
ejected from their planet of origin by electrostatic forces and then
propelled through interstellar space by the radiation pressure of
the stars. Once in space a spore might become attached to a dust
particle, which, by increasing its effective mass, would overcome
radiation pressure and cause it to fall toward the nearest star, perhaps to be captured by a planet of that star. In this way, living matter would hop from planet to planet and from one solar system to
another. As Arrhenius pointed out, one consequence of this theory
was that all living things in the universe should be chemically related to one another.

The panspermia theory rests on two propositions that should be
considered separately. The first is that life has always existed, that
it is coeternal with matter. We can say now that this thought is almost certainly wrong. Life is not one of the fundamental categories
of the universe, like matter and energy; it is, rather, a manifestation of certain molecular combinations. These combinations cannot have existed forever because even the elements of which they
are composed have not existed forever. Cosmologists tell us that
the universe consisted originally of hydrogen, the lightest element,
or of neutrons, fundamental particles of the same mass as hydrogen. All of the elements heavier than hydrogen have been produced—and are still being produced—from hydrogen by nuclear
fusion reactions in the stars. These same reactions are the source of
stellar energy. Although much hydrogen has been consumed in the
estimated 10-to-15-billion-year life of the universe, hydrogen is
still the most abundant element by far. About 90 percent of the

atoms of the visible universe—over 60 percent of the mass—consist of hydrogen, and most of the rest are helium, the second-lightest element. But since elements other than hydrogen are essential to the constitution of living matter, it follows that life cannot be as old as the cosmos, but must have had a more recent origin.

The second proposition of the panspermia theory, that spores can and do journey through the cosmos, appears much less plausible today than it appeared to Arrhenius. The effects of radiation —ultraviolet, X ray, and cosmic—that organisms would encounter in space are far more apt to be lethal than Arrhenius supposed, and interstellar distances and therefore times required for transit are much greater. We now also have empirical evidence that spores capable of seeding the universe neither leave the earth nor enter its vicinity. Samples returned from the moon by the Apollo astronauts contained no detectable microorganisms. The moon would be expected to trap a significant number of particles leaving the earth as well as those arriving from elsewhere. Biological tests of lunar surface material would reveal any organisms capable of surviving long cosmic voyages, but so far, all such tests have been negative. Spores would also have been landing on Mars ever since the solar system originated 4.5 billion (4500 million) years ago, and we shall examine the evidence for the presence of life on Mars later in this book.

Yet despite the evidence against it, the panspermia theory lives on. In recent years, the well-known astrophysicist and writer of science fiction Fred Hoyle and his collaborator, Chandra Wickramasinghe, have arrived at the implausible conclusion that at least 80 percent of interstellar dust grains consist of bacterial and algal cells. They base this identification on the optical properties of the interstellar grains. The mass of these grains in our galaxy is estimated to equal the mass of 5 million of our suns. This view therefore implies that the earth is almost lifeless compared to interstellar space. Along with Arrhenius, Hoyle and Wickramasinghe see these cells as planet hoppers. If this were true, these travelers would, of course, be arriving on the moon and on Mars.

Several scientists have recently proposed yet a newer version of the panspermia hypothesis. According to their scenario, life on the earth came from elsewhere in the cosmos, but not accidentally, as the classical panspermia theory assumes. Rather, it was brought, or sent, in interstellar spacecraft by intelligent beings living on a planet of another star. This theory does not assume that life has always existed, as Helmholtz, Kelvin, and Arrhenius did, but that it originated by chemical means (we shall examine such means in Chapter 3). The conditions necessary for an origin of life not being fulfilled on the primitive earth, however, the life now on our planet originated elsewhere in the galaxy where these conditions were satisfied. The most detailed of these proposals has been called directed panspermia by its authors, Francis Crick and Leslie Orgel. Crick and Orgel argue that enough time has elapsed since the universe began for an earlier technological civilization to have evolved in the galaxy and, for obscure reasons, to have deliberately seeded the earth some 4 billion years ago with microbial cells carried in an unmanned spacecraft.

At first I thought this hypothesis was a spoof, created to show the incompleteness of our understanding of the origin of life. When Crick published a book presenting directed panspermia as a serious alternative to the view that the life on our planet originated here (see the Bibliography at the end of the book), I changed my mind. Although no evidence favors this theory over the conventional one, none now available disproves it either. Discovery of life on another planet of our galaxy would permit one test of the hypothesis because all panspermia hypotheses predict essential identity of genetic systems, whereas the hypothesis of local origins does not.

The directed panspermia theory is part of a wider discussion currently taking place on the possibility of extraterrestrial civilizations in our galaxy. Theoretical studies of this question, as well as actual searches for radio signals from such civilizations are the subjects of increasing efforts. Although the uncertainties are still considerable, there has been a perceptible shift in recent years away from the easy assumption of the early space age that the galaxy is swarming with technologically advanced societies living on earthlike planets of other stars. Both theoretical arguments and recent explorations of our own solar system indicate that suitable planets are probably rare. Other considerations suggest that once a civilization has acquired the capability of interstellar travel it would expand rapidly (on the geological time scale) to occupy the entire galaxy. If there are ancient extraterrestrial civilizations capable of space travel, where are they? Obviously, we do not observe their presence in the solar system. (Since human beings evolved on this planet, we do not represent such civilizations.) These fascinating matters are more fully detailed in the collection of papers edited by Hart and Zuckerman (see the Bibliography).

Perhaps the surest way to wisdom lies in continued efforts to achieve an understanding of conditions on the primitive earth and to discover at least one plausible route to the spontaneous assembly of an elementary genetic system. Our progress toward this goal is the subject of Chapter 3.

CHAPTER THREE

The Origin of Life: Chemical Evolution

Out of these nothings
—All beginnings come.
Theodore Roethke, *The Longing*

Chemical evolution, the modern theory of the origin of life, also rests on the idea of spontaneous generation. It starts not with the sudden appearance of living creatures on the earth de novo, however, but with the origin of the chemical compounds and systems that make up living matter. It deals with the chemistry of the earliest earth, especially with chemical reactions in the primitive atmosphere and surface waters where the light elements that compose living matter would have been concentrated and where solar energy would have been abundant. It asks the question: How could organic compounds have arisen spontaneously in that remote era and been assembled into a living system?

THE OPARIN-UREY THEORY

The Russian biochemist A. I. Oparin (1894–1980) first stated the general approach of chemical evolution in a small book published in Russian in 1924, enlarged in 1936, and translated into English in 1938. Oparin pointed out that present conditions on the earth's surface prevent the large-scale synthesis of organic compounds because the abundance of free oxygen in the atmosphere oxidizes carbon compounds to carbon dioxide (CO_2). Besides this, he noted that any organic matter exposed on the earth today is consumed by living organisms (a point that Charles Darwin had made earlier). But, Oparin argued, conditions would have been different on the primitive earth, which could be expected to have had an atmosphere free of oxygen but rich in hydrogen or hydrogen-containing gases such as methane (CH_4) and ammonia (NH_3). (Such an atmosphere is said to be reducing—that is, hydrogen-rich and oxygen-poor—in contrast to the present oxidizing—or oxygen-rich, hydrogen-poor atmosphere.) Oparin pointed out the excellent possibilities for spontaneous synthesis of organic compounds under such conditions.

To support his idea that the earth's primitive atmosphere was reducing, Oparin cited the following evidence:

1. Hydrogen is abundant in the stars (see Figure 3-1 and Plate 1).

2. Carbon appears in the spectra of comets and cooler stars as the radicals CH and CN, but oxidized carbon is rare.

3. Hydrocarbons—that is, compounds of carbon and hydrogen—are found in meteorites.

4. The atmospheres of Jupiter and Saturn contain large amounts of methane and ammonia.

These four points, Oparin argued, show that the universe as a whole is in a reduced state. The carbon and nitrogen of the primordial earth would therefore be in the same state.

5. Ammonia is found in volcanic gases. This evidence indicated to Oparin that nitrogen was present as ammonia in the early atmosphere.

6. The oxygen of the present atmosphere is produced by green-plant photosynthesis and is therefore biological in origin.

Oparin concluded that carbon first appeared on the primitive earth as hydrocarbons and that nitrogen first appeared as ammonia. From these beginnings, he proposed, known chemical reac-

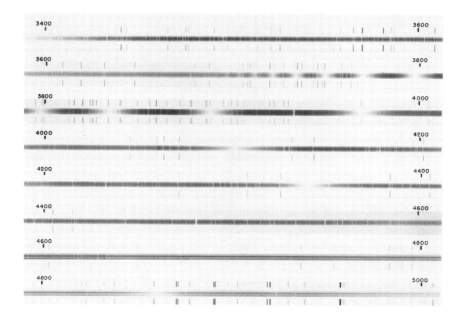

Figure 3-1 **Hydrogen lines in the spectrum of the bright star Sirius. The spectrum of the star (shown as white lines on a dark background) is paralleled by two laboratory reference spectra (dark lines on a light background). All of the very bright, broad lines in the stellar spectrum are hydrogen lines (Palomar Observatory Photograph).**

tions at the surface of the lifeless earth would have generated complex organic compounds, which, after a suitable and presumably lengthy period of time, would have given rise to the first living thing. The first organisms would have been simple systems just capable of duplicating themselves at the expense of the organic medium from which they had arisen. In modern terminology, they were "heterotrophic"—that is, dependent on the environment for their supplies of organic nourishment. At the opposite end of the scale are "autotrophic" organisms, such as green plants, which produce all of their organic substance from carbon dioxide, inorganic nitrogen, and water. On the Oparin theory, autotrophs evolved only after heterotrophs had exhausted the organic compounds in the primitive sea.

J. B. S. Haldane (1892–1964) advanced an idea similar in some respects to Oparin's in a popular essay published in 1929. Haldane proposed that organic matter, synthesized by natural processes on the prebiological earth, accumulated in the ocean, which eventually reached the consistency of a "hot dilute soup." He pointed out that the primitive atmosphere of the earth must have been anaerobic (free of oxygen), but he did not recognize that conditions had to be reducing to accomplish the synthesis of organic compounds. Thus he assumed that carbon was present in its fully oxidized form, carbon dioxide, rather than as methane or other hydrocarbons. He cited experiments (not his own) claiming to show that when a mixture of carbon dioxide, ammonia, and water is irradiated with ultraviolet light, complex organic compounds are produced. Later attempts to reproduce these experiments were unsuccessful.

In 1952, Harold Urey (1893–1981) independently arrived at the conclusion that the young earth's atmosphere had been in a reduced state. Urey was investigating the evolution of the solar system, not the origin of life specifically. Oparin's discussion had been qualitative. Urey's problem was one of physical chemistry: starting with a dust cloud of cosmic composition and the boundary conditions imposed by the known physical and chemical properties of the moon and planets, his objective was to derive a thermodynamically reasonable history of the whole system. Among other results, Urey showed that the earth, at the end of its formation, had a strongly reducing atmosphere, its major constituents being hydrogen and the fully reduced forms of carbon, nitrogen, and oxygen —that is, methane, ammonia, and water vapor. The earth's gravitational field cannot hold hydrogen, and this gas gradually escaped into space. A secondary effect of the loss of free hydrogen was the gradual oxidation of methane to carbon dioxide and of ammonia to nitrogen gas, which transformed the atmosphere from a reducing to an oxidized one over a period of time. It was during the era of hydrogen escape, while the atmosphere was in an intermediate state of oxidation-reduction, that Urey envisioned a large-scale formation of complex organic matter on the earth. The oceans, he estimated, might have become a 1 percent solution of organic compounds. The first life followed.

Urey's theory had an important consequence: it led to a successful experimental test. Before we consider experiments based on the hypothesis of a hydrogen-rich primordial atmosphere, however, we must ask how well the geological evidence supports this hypothesis. The question has been much discussed in recent years, and many geologists now doubt that a strongly reducing atmosphere ever existed on the earth. These arguments, with some modifications, are also relevant for Mars, and so it is useful to summarize them here.

THE PRIMITIVE EARTH

The solar system is believed to have formed from a great cloud of gas and dust, the solar nebula. The age of the system, based on a number of independent determinations, is close to 4.5 billion years. The best clue to the composition of the primordial nebula is found in the abundances of chemical elements presently in the solar system. The nine most abundant elements, as determined by spectroscopic studies of the sun (which contains 99.9 percent of the mass of the entire system) and, for certain elements, by the analysis of meteorites, are shown in Table 3-1. Hydrogen and helium are by far the most important constituents. Together these two gases make up over 98 percent of the mass of the sun (99.9 percent of its atoms) and ipso facto of the solar system as a whole. Since the sun is a typical star—and this category includes stars of other galaxies—its composition describes the cosmos generally. From what we know of the evolution of stars, we can say that hydrogen and helium were also predominant in the sun 4.5 billion years ago when it was new.

Table 3-1 also shows the bulk composition of the earth. Even though the four major elements of the earth are among the nine most abundant of the sun, the makeup of our planet deviates considerably from that of the cosmos as a whole. (The same is true of Mercury, Venus, and Mars but not of Jupiter, Saturn, Uranus, and Neptune.) The earth is composed mainly of iron, oxygen, silicon, and magnesium. It is deficient in all the biologically important light elements except oxygen, and it is strikingly deficient in the so-called rare, or noble, gases, such as helium and neon. On the whole, our planet seems a very unpromising place for the origin of any life.

Table 3-1   **Elemental Compositions (Mass Percent)**

Order of abundance	SOLAR SYSTEM*		EARTH†	
	Element	Percent	Element	Percent
1	Hydrogen	77	Iron	34.6
2	Helium	21	Oxygen	29.5
3	Oxygen	0.83	Silicon	15.2
4	Carbon	0.34	Magnesium	12.7
5	Neon	0.17	Nickel	2.4
6	Nitrogen	0.12	Sulfur	1.9
7	Iron	0.11	Calcium	1.1
8	Silicon	0.07	Aluminum	1.1
9	Magnesium	0.06	Sodium	0.57
	Total	**99.70**	Hydrogen + carbon + nitrogen	0.05
			Neon	1×10^{-9}
			Total	**99.12**

* After Cameron (1970).
† After Mason (1966).

The basic Oparin-Urey position held that the atmosphere of the young earth, reflecting the chemistry of the solar nebula, was strongly reducing. However that may be, the earth's atmosphere is now oxidizing. The present atmosphere contains 77 percent nitrogen, 21 percent oxygen, 1 percent water vapor (on average), nearly 1 percent argon, and traces of other gases. How, then, might a reducing atmosphere have originated? Perhaps a large endowment of the gases of the solar nebula provided the newly forming earth at the outset with the hydrogen and other light elements that were needed to initiate an Oparin-Urey evolution. Reasonable though this possibility may seem, the deficiency of light elements and especially of the noble gases shows that the earth was formed without an atmosphere. With the exception of helium, all of the noble gases—neon, argon, krypton, and xenon—are heavy enough to be held by the earth's gravitational field. Krypton and xenon are heavier than iron. Because these elements make very few compounds, they would have existed largely as gases in the atmosphere of the primitive earth, and once the earth attained its present size, they could not have been lost. Their deficiency by factors of millions in the terrestrial inventory argues that the earth never had an atmosphere of solar composition but was formed out of solid materials that, except for small amounts of occluded or adsorbed gas, had no atmosphere associated with them. The elements that now form our atmosphere must have arrived on the primordial earth as solid chemical compounds that were later decomposed into gases by the heat generated by radioactivity and by the release of gravitational energy that accompanied accretion of the earth. These gases were exhaled from the earth's interior through volcanoes to form the primitive atmosphere.

The high argon content of the present atmosphere (almost 1 percent) does not contradict the conclusion that the noble gases were primordially absent. The cosmically abundant form of argon has an atomic weight of 36, while our argon, which has been generated by the radioactive decay of potassium in the earth's crust, has an atomic weight of 40. The fact that oxygen combines with a variety of other elements to form very stable, solid compounds, such as the silicates and carbonates that compose the rocks, explains the anomalously high abundance of oxygen in the earth relative to the other light elements.

Urey's argument for the reducing character of the primordial atmosphere was based on the high iron content of the earth (35 percent by mass). He assumed that this iron—which now forms the core of the earth—was originally distributed more or less uniformly throughout the planet. When the earth heated up, the iron melted and flowed into the center. Before that happened, however, the iron in what is now the earth's upper mantle reacted with water (which had arrived on the primitive earth in the form of hydrated minerals like those that occur in certain meteorites), and this released large amounts of hydrogen into the primordial atmosphere.

Studies since the early 1950s have called some parts of this scenario into question. Some planetary scientists consider it unlikely that the iron now in the earth's core was ever distributed homogeneously throughout its bulk. They favor an inhomogeneous mode of accretion in which iron condensed out of the nebula earlier than most of the materials that now form the mantle and crust. This would mean less free hydrogen in the primitive atmosphere than would be expected if accretion were homogeneous. Others favor some manner of homogeneous accretion that would not necessarily result in a reducing atmosphere. In short, a variety of different planet formation models have been considered in recent years, some of them more and some less accommodating to the idea of an early reducing atmosphere.

Attempts to reconstruct events that occurred in the remotest ages of the solar system are, of course, subject to very great uncertainties. The gap in the geological record between the origin of the earth and the origin of its oldest dated rocks—the period during which the chemical events that led to the origin of life took place —is 700 million years. Laboratory experiments have shown that synthesis of the components of the genetic system requires reducing conditions, and we can therefore say that if life originated on the earth, either the primitive atmosphere was reducing or the organic compounds required for life to begin arrived from elsewhere. Since even today meteorites bring to earth a variety of organic materials, including many compounds of biological interest, the latter possibility is not altogether fanciful. Meteorites do not appear to contain everything needed to construct a genetic system, however. Although meteorite-borne materials may well have contributed to the pool of organic matter on the primitive earth, the most plausible view at this time holds that the earth itself was sufficiently reducing to allow the production of the organic matter that led to the origin of life.

EXPERIMENTS ON PREBIOTIC CHEMISTRY: SYNTHESIS OF MONOMERS

Oparin apparently never attempted an experimental test of his theory. Perhaps he felt that the analytical methods available were inadequate to characterize the complex mixtures that would be expected from the many reactions possible among hydrocarbons, ammonia, and water. Or perhaps he was satisfied with the general principles revealed by logic and did not care to get into the messy details. Whatever the explanation, the Oparin theory was never tested until Urey rediscovered it. Then Stanley Miller, a student of Urey's, carried out a famous experiment that transformed the origin-of-life problem from a matter for speculation into a branch of experimental chemistry.

Miller passed an electric discharge through a mixture of the four gases of Urey's proposed atmosphere—hydrogen, methane, ammonia, and water vapor—in the presence of liquid water. At the end of a week, he analyzed the dissolved products and found a significant number of substances of biological interest, including glycine, alanine, aspartic acid, and glutamic acid, four of the amino acids found in proteins. The experiment was later repeated using better analytical methods and a gas mixture more consistent with presently favored models of the primitive atmosphere. Nitrogen replaced most of the ammonia—the latter would have dissolved in the primitive ocean—and hydrogen, now thought to have been at best a minor component of the primitive atmosphere, was omitted. This experiment produced 12 of the amino acids found in proteins[1], along with many others unrelated to proteins but of no less interest for reasons about to be explained.

Study of these remarkable syntheses showed that the electric spark generates certain primary products that in turn undergo further reactions when they dissolve in water to yield the final products. The most important primary products are hydrogen cyanide (HCN), formaldehyde (HCHO) and other aldehydes, and cyanoacetylene (HCCCN). Amino acids are formed from hydrogen cyanide in at least two different ways: one involves the reaction of cyanide, aldehydes, and ammonia in solution; the other involves the transformation of HCN alone into amino acids through a complex series of reactions in solution.

---

[1] The amino acids were glycine, alanine, valine, leucine, isoleucine, proline, aspartic acid, glutamic acid, serine, threonine, asparagine, and glutamine.

It seems likely that then, as now, solar radiation, not electric discharge, was the major source of energy reaching the earth. Accordingly, ultraviolet light has been tested by various investigators as a source of energy for amino acid synthesis, with positive results. The highest yields are obtained when hydrogen sulfide (H_2S), which absorbs the longer, more abundant, ultraviolet wavelengths, is incorporated into the Urey gas mixture. Shock waves that generate brief pulses of high temperature and pressure have also been found to produce amino acids. Mechanical energy such as this would have been available in the primitive ocean from wave action and in the atmosphere from thunder and meteorites.

Important extensions of the Miller experiment by Juan Oró and by Leslie Orgel and his associates have shown that the four bases of RNA (three of which also occur in DNA) are formed in other reactions derived from the primary products of the spark discharge. Specifically, hydrogen cyanide condenses with itself in a series of reactions in solution to yield the purine base adenine; a variation of these reactions produces the other purine, guanine. The pyrimidine bases cytosine and uracil are obtained in good yields from cyanoacetylene in reactions that also could have taken place on the primitive earth. A prebiotic synthesis of thymine, which is found in DNA in place of uracil, has not yet been reported, however.

Finally, it has been known for a long time that formaldehyde condenses in solution under certain conditions to yield a variety of sugars. Ribose, the sugar component of RNA, is one of the products of this reaction. Thus we now know that most of the molecular components of the genetic system can be formed by means of reactions that are, at the least, plausible for the primitive earth.

METEORITES AND INTERSTELLAR CLOUDS

Recent discoveries in the chemistry of meteorites and interstellar dust clouds show that the synthesis of biologically important molecules has occurred in our galaxy on a vast scale and is still taking place. The meteorites in question are the carbonaceous chondrites, a class that makes up approximately 5 percent of the meteorites that fall on the earth every year. Considered to be primitive, largely unmodified rubble left over from the solar nebula, these interesting objects date from the time of formation of the solar system 4.5 billion years ago. They are much too small to possess atmospheres, but in their abundances of nonvolatile elements carbonaceous chondrites closely resemble the sun. The minerals that compose them indicate that they were formed at a low temperature and have never been exposed to high temperatures. They contain up to 20 percent water (combined as mineral hydrates) and up to 10 percent organic matter.

Carbonaceous chondrites have attracted attention for their possible biological significance since the nineteenth century. Upon discovering organic matter in the Alais meteorite, which fell in France in 1806, the Swedish chemist Jakob Berzelius raised the question whether this indicated the existence of extraterrestrial life. He thought not. Pasteur is said to have had a special probe constructed in order to obtain an uncontaminated sample from the interior of the Orgueil meteorite, another famous chondrite that fell in France in 1864. He tested the sample for the presence of microorganisms, with negative results.

Until recently, little significance could be attached to the identification of organic compounds in carbonaceous chondrites because of the difficulty in distinguishing between indigenous substances and contaminants acquired as a meteorite enters the earth's atmosphere, hits the ground, and is subsequently handled. This has now changed, thanks to ultrasensitive analytical methods and to the recognition that rapid recovery and clean handling of observed falls is necessary. Two recently fallen chondrites—the Murchison, which fell in Australia in 1969, and the Murray, which fell in the United States in 1950—have both contained indigenous amino acids.[2] Several lines of evidence show that these amino acids are not contaminants. For one, many of the meteoritic amino acids are unusual kinds that do not exist in terrestrial organisms. For another,

---

[2] Over 50 amino acids have been identified in the Murchison meteorite, eight of which are found in proteins: glycine, alanine, valine, leucine, isoleucine, proline, aspartic acid, glutamic acid. Two additional protein amino acids, serine and threonine, have been found, but these may be contaminants.

some common amino acids that are expected to show up in contaminated samples do not appear in the meteorites. Finally, the amino acids in these meteorites occur in both their mirror-image, or optical isomeric, forms, a phenomenon expected of nonbiologically synthesized amino acids but not of those in living organisms (as we saw in Chapter 1).

The sets of amino acids found in meteorites show a marked resemblance to those produced in the spark-discharge experiment. The two sets are not identical, but the similarity is strong enough to suggest that some of the same synthesizing mechanisms operate in both cases. Another proposed mechanism for the synthesis of amino acids in meteorites is the Fischer-Tropsch reaction, named after two German chemists who invented a catalytic process for producing gasoline and other hydrocarbons from carbon monoxide (CO) and hydrogen. Both of these gases are plentiful in the universe, as are suitable catalysts—iron, or silicates, for example. In investigating the possibility that this process explains the cosmic abundances of organic compounds, Edward Anders and his co-workers at the University of Chicago have found that amino acids, purines, and pyrimidines are generated if ammonia is incorporated in the reaction mixture. These syntheses may involve the same intermediates—hydrogen cyanide, aldehydes, cyanoacetylene—as those found in the electric-discharge reaction. Fischer-Tropsch-type syntheses appear to explain the presence of hydrocarbons as well as purines and pyrimidines in meteorites more readily than the electric-discharge reaction does. None of the laboratory experiments, however, has yet duplicated in detail the catalogue of compounds found in meteorites.

Less work has been done on the purine and pyrimidine bases of meteorites than on their amino acids, but adenine, guanine, and uracil have all been reported in the Murchison meteorite. Adenine and guanine were detected at levels around one to 10 parts per million, similar to the reported abundances of amino acids. The uracil level was much lower.

The recent discovery by radioastronomers of organic molecules in interstellar space has added a new dimension to our view of the organic chemistry of the universe. The molecules in question exist in association with the great clouds of gas and dust that are found in regions of the sky where stars and planetary systems are believed to be forming. Nearly 60 compounds in addition to the expected hydrogen molecules have been identified at this writing. The most abundant of these is carbon monoxide. Less abundant but equally interesting are ammonia, hydrogen cyanide, formaldehyde, acetaldehyde (CH_3CHO), cyanoacetylene, and water—all known to be precursors of amino acids, purines, pyrimidines, and sugars in laboratory experiments on chemical evolution.

These discoveries show that the synthesis of organic matter occurs widely in the universe and that among the products are many compounds of biological interest, including the basic subunits of the genetic system and their precursors. It is even possible, as suggested earlier, that the organic compounds, or some part of them, that went to form the first living organism were extraterrestrial in origin. The findings have led to the important realization that the synthesis of biological compounds is not some special kind of chemistry that is restricted to particularly favorable environments, such as our planet, but is cosmic in its range. This immediately suggests that wherever life occurs in the universe it will be based on carbon chemistry similar to, although not necessarily identical with, our own.

PREBIOTIC SYNTHESIS OF POLYMERS

Formation of the fundamental subunits of proteins and nucleic acids from gases of the solar nebula is only the start of the beginning of the assembly of a genetic system. The subunits, or monomers, must next be joined in chains to make useful polymers. This is a difficult problem, and although it has attracted much attention, it is fair to say that a convincing route to genetic polymers, starting from the monomers that were probably available on the primitive earth, has yet to be demonstrated.

Polymer synthesis in living systems and in the laboratory involves the stepwise addition of subunits to the end of a growing chain. At each step, energy is required, and a molecule of water is eliminated. In the case of protein synthesis from amino acids, the link is called a peptide bond. In this illustration of peptide bond formation between two amino acid molecules,

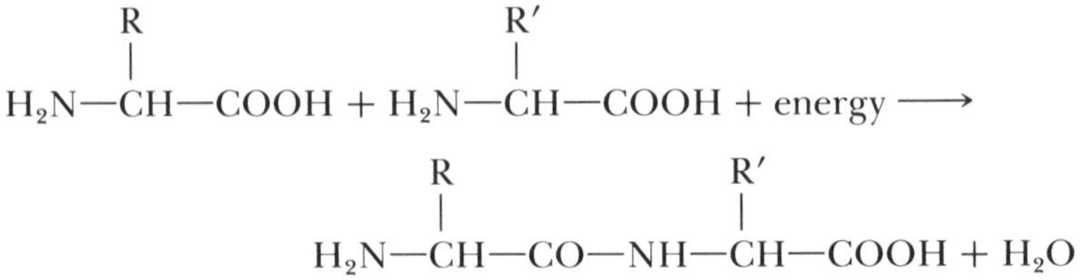

$$
\begin{array}{cc}
\text{R} & \text{R}' \\
| & | \\
\text{H}_2\text{N—CH—COOH} + \text{H}_2\text{N—CH—COOH} + \text{energy} \longrightarrow
\end{array}
$$

$$
\begin{array}{cc}
\text{R} & \text{R}' \\
| & | \\
\text{H}_2\text{N—CH—CO—NH—CH—COOH} + \text{H}_2\text{O} \quad (3\text{-}1)
\end{array}
$$

the R's stand for any of the 20 different side chains of the amino acids found in proteins. When a third amino acid molecule is attached to the end of the dipeptide, a tripeptide results, and so on until a polypeptide is formed. The reactions are reversible: the dipeptide shown above, for example, could take up a water molecule and revert to amino acids, with a release of energy. A protein molecule is a polypeptide chain with a particular sequence of amino acids that gives the molecule its special properties and that is the product of a long evolution. Each chain is hundreds of amino acids in length, and some proteins consist of two or more such chains. As a result of interactions among its constituent amino acids, the polypeptide(s) folds up into a three-dimensional structure, and this structure is the active form of the protein molecule.

The polymerization of nucleotides, the repeating subunits of nucleic acids, yields polynucleotides or nucleic acids. The formation of a dinucleotide from two nucleotides looks like this:

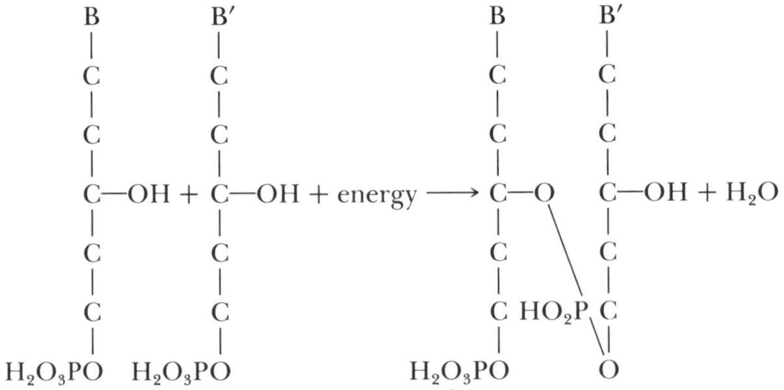

$$
\begin{array}{cccc}
\text{B} & \text{B}' & \text{B} & \text{B}' \\
| & | & | & | \\
\text{C} & \text{C} & \text{C} & \text{C} \\
| & | & | & | \\
\text{C} & \text{C} & \text{C} & \text{C} \\
| & | & | & | \\
\text{C—OH} + & \text{C—OH} + \text{energy} \longrightarrow & \text{C—O} & \text{C—OH} + \text{H}_2\text{O} \\
| & | & | \diagdown & | \\
\text{C} & \text{C} & \text{C} \diagdown & \text{C} \\
| & | & | \diagdown & | \\
\text{C} & \text{C} & \text{C} \ \text{HO}_2\text{P} \diagdown \text{C} \\
| & | & | & \| \\
\text{H}_2\text{O}_3\text{PO} & \text{H}_2\text{O}_3\text{PO} & \text{H}_2\text{O}_3\text{PO} & \text{O} \qquad (3\text{-}2)
\end{array}
$$

Here the B's stand for any of the four bases of DNA or of RNA; the chains of C's are the five-carbon sugar, with the –OH group attached to carbon atom number 3 shown. (The sugars are shown in their actual ring structures in Figure 1–1.) Phosphoric acid is combined at carbon atom number 5 at first, then at carbons 5 and 3.

To accomplish polymer synthesis—either protein or nucleic acid—living cells produce energy-rich molecules which, through the mediation of specific proteins (enzymes), make energy available for each step of monomer addition. Besides making pathways available for these reactions, enzymes provide an environment for the reaction that excludes all interfering molecules—an essential function when useful molecules form only a small fraction of those present. Water, which invariably interferes with dehydration reactions, is among the molecules excluded.

Biological polymers can also be synthesized in the laboratory without using enzymes. In fact, machines now routinely carry out polypeptide and polynucleotide syntheses. Proteins identical to those produced by cells can be and have been made in the laboratory. These procedures make use of nonaqueous solvents, purified monomers in high concentrations, and a variety of sophisticated protecting and energy-yielding reagents that, in effect, accomplish the functions normally performed by enzymes.

Contrast these two highly refined ways of biopolymer synthesis, that of the cell and that of the laboratory, with the state of affairs on the primitive earth. There water was the only solvent available, the relevant monomers made up just a fraction of the total organic and inorganic matter in solution, the only reagents present in any abundance were presumably rather simple ones, and, of course, there were no enzymes. The problem of how even short polymers might form under such unfavorable conditions is as yet unsolved. The primordial soup was, presumably, composed of a great variety of organic compounds. In order to form a polypeptide or a polynucleotide, a particular group of compounds must emerge from the soup, come together, and combine with one another. The first step is perhaps the most difficult to imagine. A mere concentration of the soup would not help much. The complex mixture would still contain many substances that would interfere with the formation of polymers—by attaching themselves to the end of the growing chain and stopping its growth, for example.

A possible solution to this problem involves adsorption of the desired molecules to the surfaces of clay minerals. This mechanism was emphasized especially by the late J. D. Bernal (1901–1971), a well-known British crystallographer. Clay minerals have large absorptive capacities for organic compounds, and they also show some specificity in the kinds of compounds they adsorb. Bernal himself expressed doubts about his proposal on the ground that silicon, the major element of these clays, has almost no role in biochemistry today. Nevertheless, adsorption remains one of the more plausible, if still unproven, prebiotic separation-concentration mechanisms.

Others have not hesitated to give silicate minerals a major role in the origin of life, despite Bernal's reservations. The chemist A. G. Cairns-Smith, of the University of Glasgow, has in fact espoused the view that life began as mineral crystals. With their ability to replicate themselves, inorganic crystals show the beginnings of genetic properties. They also have a limited capacity for mutation—that is, they pass along imperfections in the regular arrangement of atoms in the crystal. Minerals such as clays that are built up of layers tend to imprint the defects of one layer onto the structure of the next one, giving a kind of genetic memory. Noting that defects in crystal faces are often sites of chemical activity, including catalysis, Cairns-Smith suggests that a simple organic compound, such as formaldehyde, whose synthesis was catalyzed by a mineral bearing such a defect had the effect of improving the rate or the accuracy of reproduction of the defective crystal, which thereby came to outnumber other kinds. From such a beginning, there followed evolution of the protein–nucleic-acid genetic system, which then abandoned its mineral ancestor. This imaginative proposal rests on almost no supporting evidence.

With all the formidable difficulties in accounting for the first appearance of the important polymers, some mitigating factors should be borne in mind. It is highly likely that the first genetic system needed only short polymers to get started, not the large, highly evolved molecules that we find in modern organisms. The first organism need not have been very efficient. Since it lived in a Garden of Eden with no enemies and no problems of food supply, it had only to reproduce itself fast enough to stay ahead of its own chemical decomposition. Furthermore, the chemical events that preceded the origin of life took place in a vast arena of space and time. For hundreds of millions of years the primitive earth was an immense laboratory where, because of the grand scale of the operations, even events that we consider very improbable might have occurred.

Such considerations do not allow us to claim that we understand how the first biopolymers were produced. They do suggest, however, that the problem may not be quite as difficult as we think. An important recent development in Orgel's laboratory has shown that, given an initial polynucleotide chain, it is possible for the chain to produce a complementary copy of itself from nucleotides in a manner analogous to gene duplication, but without the intervention of an enzyme. This remarkable result was made possible by the discovery of a means of supplying energy to the reaction that, while not dependent on enzymes, resembles the natural mechanism that cells use to supply energy for polynucleotide synthesis. This demonstration makes plausible the idea that a similar process was important early in the evolution of the genetic system. Along with other recent evidence showing that some RNAs possess catalytic properties of the sort usually associated with proteins, it suggests that a primitive genetic system might be built up from RNA alone, without proteins. If this is so, it would greatly simplify the origin-of-life puzzle.

The problems of explaining the source of the first nucleic acid strand, and the genetic code, and the whole mechanism of information transfer from nucleic acids to proteins still remain, but we have come as far as present knowledge can bring us. This brief review of ideas about the nature and origin of life on our planet therefore ends without a solution to the origin of Pooh-Bah's ancestral "protoplasmal primordial atomic globule." We can be confident that progress toward this goal will continue. In the meantime, the principles developed in these three chapters are sufficiently general to apply even to life elsewhere in the universe, and we can now turn to the question of life on other planets of our solar system, the subject of the remaining chapters of this book.

CHAPTER FOUR

Are the Planets Habitable?

However, most of the planets are certainly inhabited, and those that are not will be someday.

Everything is summarized in one general concept: The material from which the inhabitants of the different planets are made, animals and plants alike, must above all be lighter and finer . . . the farther their habitat is from the sun.

The excellence of the intelligent creatures, the speed of their thought . . . become more perfect and complete the farther their habitation is from the sun.
   Since this relation has a credibility not far removed from established fact, we are free to make conjectures about the nature of these various inhabitants.

Immanuel Kant, *Allgemeine Naturgeschichte und Theorie des Himmels* (1755)*

* Author's translation.

In the seventeenth and eighteenth centuries, the known planets were widely believed to be inhabited. Christian Huygens (1629–1695), one of the founders of modern astronomy, considered it reasonable to suppose that Mercury, Venus, Mars, Jupiter, and Saturn possessed fields, "warm'd by the kindly Heat of the Sun, and water'd with fruitful Dews and Showers." These fields, he thought, were home to plants and animals. If it were otherwise, he wrote, these planets "would be inferior to our Earth," a condition that he found totally unacceptable. This argument, so strange sounding to us, sprang from the Copernican world view, which Huygens took as a given, that the earth holds no privileged place among the planets. For the same reason, Huygens argued, the planets must also have intelligent inhabitants, "not Men perhaps like ours, but some Creatures or other endowed with Reason." He judged this conclusion to be so certain that he wrote, "If I am mistaken in this, I do not know when to trust my Reason, and must allow my self to be but a poor Judge in the true Estimate of Things."

Although Huygens erred in this judgment—the other planets eventually proved to be far inferior to the earth, at least as habitats for life—his reputation has not suffered. His genius was wide-ranging, and his discoveries in mathematics, mechanics, astronomy, and optics are fundamental to modern science. The lesson in this for us is that where extraterrestrial life is concerned, even the most talented among us can go off the track.

Little had changed a century later, as the epigraph heading this chapter shows. Immanuel Kant was convinced not only that the planets can and do support life, he also believed that the quality of their inhabitants increases with the planet's distance from the sun.

Of course, little was known about the planets in the seventeenth and eighteenth centuries, and even less about the nature of life. At about the same time that Huygens was putting his reason on the line in defense of extraterrestrial life, Francesco Redi was proving that animals are not spontaneously generated, and thereby taking the first step toward a fundamental understanding of the nature of living things. It would be a long time before biologists and planetary astronomers would be able to assess the habitability of the planets realistically. But by the time of the Viking launch in 1975, as we shall see in this chapter and the following one, of all the planets known to Huygens, only Mars remained as a possible host of life.

CRITERIA FOR HABITABLE PLANETS

Temperature and Pressure

If our conclusion that life must be based on the chemistry of carbon is correct, then we can specify some definite bounds for any life-sustaining environment. For one thing, the temperature must not exceed the stability limit of organic molecules. This temperature is not easily defined, but our purpose does not require a precise figure. Since temperature effects and pressure are interdependent, however, we must consider both together. Given pressures in the neighborhood of one atmosphere, as on the earth's surface, we can estimate the upper temperature limit for life using the fact that many of the small molecules that make up the genetic system—amino acids, for example—decompose rapidly at 200 to 300 degrees Celsius. From this, it is probably safe to conclude that environments whose temperatures exceed 250°C are not habitable. (This is not to say that all life depends on amino acids; we are only taking these substances as representative of small organic molecules.) The actual temperature limit would almost certainly be lower than this, since large molecules with elaborate three-dimensional structures—such as proteins built from amino acids—are typically more sensitive to heat than small molecules. For life on the earth's surface, the upper temperature limit is close to 100°C, where a number of bacterial species found in hot springs can grow. The vast majority of organisms, however, die at this temperature.

It seems an odd coincidence that the upper temperature limit for life is close to the boiling point of water. Might this limit be set by the fact that liquid water cannot exist at temperatures above its boiling point (100°C on the earth's surface), rather than by any properties inherent in living matter?

Years ago, Thomas D. Brock, a specialist in high-temperature bacteria, suggested that life would be found wherever liquid water exists, regardless of its temperature. Now in order to raise the boiling point of water, one has to increase the pressure, as in a pressure cooker. Merely turning up the heat makes water boil faster without changing its temperature. Natural environments where liquid water exists at temperatures above its normal boiling point are found in submarine geothermal areas where superheated water is ejected from within the earth under the combined pressures of the atmosphere and the overlying ocean. In 1982, K. O. Stetter discovered bacteria from such an area, with depths of up to 10 meters, that grew optimally at 105°C. Since the pressure of 10 meters of water is equal to 1 atmosphere, the total pressure at this depth was 2 atmospheres. The boiling point of water under 2 atmospheres pressure is 121°C. Water temperatures of 103°C were actually measured at this site. Life is thus possible at temperatures above the normal boiling point of water.[1]

Obviously, bacteria that live at temperatures in the neighborhood of 100°C possess a secret not shared by ordinary organisms. Since these high-temperature forms grow poorly, if at all, at low temperatures, it is equally true that ordinary bacteria have their own secret. The key to survival at high temperatures is the ability to produce heat-resistant cellular components—especially proteins, nucleic acids, and cell membranes. The proteins of ordinary organisms undergo a rapid, irreversible change of structure, or denaturation—the coagulation of egg albumen (egg "white") by cooking, for example—at temperatures around 60°C. Proteins from hot-spring bacteria do not undergo such changes until the temperature is 90°C. Nucleic acids also undergo heat denaturation. In the case of DNA, denaturation results in the separation of its two strands. Typically, this occurs in the temperature range between 85 and 100°C, depending on the specific composition of the DNA.

---

[1] A recent report that bacteria collected from geothermal vents on the bottom of the Pacific Ocean (see page 61) can grow at 250°C under a pressure of 265 atmospheres has been seriously questioned and is probably erroneous.

In proteins, denaturation destroys the three-dimensional configuration, unique for each protein, that is required for the performance of functions such as catalysis. This configuration is maintained by a variety of weak chemical bonds that cause the linear array of amino acids forming the basic structure of the protein molecule to fold up into the specific conformation that characterizes the protein. The bonds that maintain the three-dimensional shape are formed between amino acids located in different parts of the molecule. Mutations in the gene that carries amino acid sequence information for a specific protein may cause changes in the amino acid composition, and this, in turn, frequently alters its thermostability. This effect makes possible the evolution of heat-stable proteins. The molecular structures that confer heat resistance on the nucleic acids and cell membranes of hot-springs bacteria are likewise genetically based.

Just as raising the pressure prevents water from boiling at its normal boiling point, so it can prevent some of the damaging effects of high temperature on biological molecules. Pressures of a few hundred atmospheres, for example, inhibit the heat-denaturation of proteins. This happens because denaturation causes an unfolding of the protein molecule, with an accompanying increase in volume. By opposing the volume increase, pressure prevents denaturation. At much higher pressures—5000 atmospheres or more —pressure becomes a denaturing agent itself. The mechanism of this effect, which suggests compressional collapse of the protein molecule, is not understood. Another effect of very high pressures is to stabilize small molecules against thermal decomposition by opposing the volume increase associated with the breaking of chemical bonds. Urea, for example, decomposes rapidly at 130°C at atmospheric pressure but is stable for at least an hour at 200°C under a pressure of 29,000 atmospheres.

For molecules in solution, the situation is quite different. These are often decomposed at high temperature by reacting with the solvent. The general name for such reactions is solvolysis; when the solvent is water, the reaction is called hydrolysis. (Reactions 3-1 and 3-2, on page 50, are typical hydrolyses when read from right to left.) Reaction 3-1, rewritten here as a hydrolysis, emphasizes the fact that amino acids in solution are ions and carry electric charges.

$$\begin{array}{ccccc} R & & & R' & \\ | & & & | & \\ {}^+H_3N-CH-CO-NH-CH-COO^- + H_2O & \longrightarrow \end{array}$$

$$\begin{array}{ccccc} & R & & & R' \\ & | & & & | \\ & {}^+H_3N-CH-COO^- + {}^+H_3N-CH-COO^- & & (4\text{-}1) \end{array}$$

Hydrolysis is the major process by which proteins, nucleic acids, and many other complex biological molecules are destroyed in nature. Food digestion in animals uses it, for example, but it also occurs spontaneously outside of living systems, especially at high temperatures. The electric fields generated by solvolytic reactions bring about a reduction in volume by electrostriction, or binding, of solvent molecules in the neighborhood. One would therefore expect solvolysis to be accelerated by high pressure, and experiment proves this to be the case.

Since we believe that living processes can take place only in systems in solution, we must therefore conclude that high pressure cannot extend the upper temperature limit for life, at least in polar solvents such as water and ammonia. The limit of about 100°C may be inescapable. This result removes many parts of the solar system from consideration as possible habitats, as we shall see.

Atmosphere

The next requirement for a habitable planet is that it have an atmosphere. Small compounds of the light elements that we assume compose living matter are frequently volatile—that is, they are gaseous—over a wide temperature range. It seems inescapable that living organisms will produce such compounds in the course of their metabolism, as will thermal and photochemical processes acting on dead organisms, and that these gases will enter the atmosphere. On the earth, such gases—carbon dioxide, water vapor, and oxygen are common examples—are eventually recycled through the living world. If the earth's gravitational field were unable to hold them, they would be lost to space, our planet would in time be depleted of its light elements, and life would cease. Thus, even if life could get started on an object incapable of retaining an atmosphere, it could not survive there indefinitely.

It has been suggested that life might exist beneath the surface of bodies such as the moon, which have no atmosphere or only a thin one. The idea is that gases may be trapped in subsurface environments that could then become suitable biological habitats. But since any subsurface habitat would be cut off from the ultimate biological energy source, the sun, this proposal only exchanges one problem for another. Life requires a constant supply of both matter and energy, but whereas matter recycles (hence the need for an atmosphere), energy does not, for fundamental thermodynamic reasons. The biosphere can function only as long as it is supplied with energy, but all energy sources are not equal. Heat, for example, is plentiful in the solar system—indeed, it is produced in the interiors of many planets, including the earth. Yet no species capable of using heat as a source of energy for life processes is known. In order to use heat as an energy source, an organism would have to function as a heat engine—that is, it would have to move heat from a place with a high temperature (for example, the cylinder of a gasoline engine) to a place with a low temperature (the radiator). In this process, a fraction of the transferred heat becomes available to do work. But reasonable efficiencies from heat engines require high temperatures, and this immediately poses great difficulties for organic systems—as does a host of additional problems.

Sunlight poses none of these problems. The sun is a constant, virtually inexhaustible source of energy, one that is readily usable for chemical purposes at any practical temperature. Life on our planet is totally dependent on solar energy, and we can assume that life elsewhere in the solar system would not survive without direct or indirect access to the sun.

The fact that some bacteria can live in the dark in a wholly inorganic environment, with carbon dioxide as their only source of carbon, does not alter this conclusion. These organisms, called *chemolithoautotrophs* (literally, inorganic-chemical self-feeders), obtain energy for converting carbon dioxide into cellular organic material by oxidizing hydrogen, sulfur, or other inorganic substances. Such energy sources, unlike the sun, are exhaustible, and once used they cannot be renewed without an input of solar energy. For example, hydrogen, an important fuel source for some chemolithoautotrophs, is formed in anaerobic environments, such as swamps, lake bottoms, and the digestive tracts of animals, by bacterial decomposition of plant materials (materials themselves photo-synthetically derived, of course). Chemolithoautotrophs use this hydrogen to produce methane and cell material from carbon diox-ide. The methane enters the atmosphere, where it is decomposed by sunlight to yield hydrogen and other products. The earth's atmosphere contains 0.5 part per million of hydrogen, almost all of it originating from microbially formed methane. The hydrogen and methane produced by volcanoes is negligible in comparison. The only other source of atmospheric hydrogen lies at the top of the atmosphere where hydrogen atoms that have been split from water vapor by the action of solar ultraviolet are found and from where they escape into space.

The dense populations of animals—fish, clams, mussels, gigantic worms, and so on—that have been discovered living around the vents of hot springs 2500 meters beneath the surface of the Pacific Ocean are often described as being independent of solar energy. Several such areas are known, one near the Galapagos Islands and another some 21 degrees to the north and west, off the coast of Mexico. Food supplies are notoriously scarce in the deep ocean, and the discovery of the first such population in 1977 immediately raised the question of its source of nourishment. One possibility appeared to be the organic matter on the ocean floor, the fallout from biological activity at the surface, that is carried into the vent areas by horizontal currents set up by the hot water spewing out of the vents. The rise of this heated water brings currents of cold water in along the bottom, and thus, it was thought, organic debris would collect around the vents.

Another source of nutrients came to light following the discov-ery that the water from these springs contains hydrogen sulfide (H_2S). This raised the possibility that chemolithoautotrophic bacte-ria are at the base of the food chain. Investigation has established that chemolithoautotrophy is indeed the principal source of or-ganic matter in the hot-springs ecosystem. The bacteria in question carry out the following reaction:

$$CO_2 + 4H_2S + O_2 \longrightarrow [CH_2O] + 4S + 3H_2O \qquad (4\text{-}2)$$

where $[CH_2O]$ stands for carbohydrate or cell material generally.

Since hydrogen sulfide, the fuel of these deep-sea communities, originates deep in the earth, these systems are often described as operating on a nonsolar energy source. This description is not quite accurate, however, because the oxygen they use for oxidizing the fuel is a photoproduct. There are only two significant sources of free oxygen on the earth, and both depend on the sun. The major source is green-plant photosynthesis (performed not only by green plants but also by some bacteria):

$$6CO_2 + 6H_2O + \text{sunlight} \longrightarrow C_6H_{12}O_6 + 6O_2 \qquad (4\text{-}3)$$

where $C_6H_{12}O_6$ is the sugar glucose. The other very minor source of free oxygen is photodecomposition of water vapor in the upper atmosphere. If a vent organism were found that operated entirely on gases originating deep in the earth, then a metabolism truly independent of solar energy would be known.

It should be realized that the ocean plays an essential role in the deep-sea ecosystem, since it provides an environment for the vent organisms that they could not do without. The ocean brings them not only oxygen but all the nutrients, except hydrogen sulfide, that they need. It carries away waste. And it enables them to colonize new sites—an activity that is essential for the survival of the vent organisms, since the vents are ephemeral with estimated lifetimes of the order of 10 years. Distances between vent areas in the same general region of the ocean are 5 to 10 kilometers.

Solvent

Everyone agrees that a solvent of some kind is essential for life. Without a solvent, the many chemical interactions that must take place in living systems would be impossible. On the earth, the biological solvent is water. Water is the principal constituent of living cells, and it is one of the most plentiful substances on the earth's surface. In view of the large cosmic abundances of the elements that compose water, it is doubtless one of the most widely distributed substances in the universe, as well. Yet despite this plenitude of water, the earth is the only planet in the solar system that has an ocean on its surface, an important fact to which we shall return later.

Water has a number of special and unexpected properties that make it suitable as a biological solvent and, beyond this, as a habitat for living organisms. Its characteristics also give it a major role in stabilizing the temperature of the earth. These properties include its high melting and boiling points, its high heat capacity, the wide temperature range over which it remains liquid, its very high dielectric constant (important for its solvent action), and its expansion near the freezing point. The classic treatment of this subject is that of L. J. Henderson (1878–1942), a professor of chemistry at Harvard.

Modern work has shown that the unusual characteristics of water result from the ability of water molecules to form hydrogen bonds with one another and with other molecules containing oxygen or nitrogen atoms. Liquid water actually consists of clumps of water molecules held together by hydrogen bonds. For this reason, speculations on nonaqueous solvents for life on other worlds usually focus on ammonia (NH_3), which also makes hydrogen bonds and shares many of the properties of water. Other hydrogen-bond formers, such as hydrofluoric acid (HF) and hydrogen cyanide (HCN), are also sometimes mentioned. The last two substances are implausible candidates for this role, however. Fluorine is not an abundant element: about 10,000 atoms of oxygen exist for every fluorine atom in the visible universe, and it is hard to imagine circumstances that would favor the accumulation of an ocean of HF rather than of H_2O on any planet. As for HCN, the elements that compose it are cosmically abundant, but HCN is thermodynamically unstable with respect both to its elements and its likely precursors. It is therefore improbable that large amounts of this substance could ever accumulate on a planet although, as we saw earlier, HCN is an important (but transient) intermediate in the nonbiological synthesis of organic compounds.

Ammonia is composed of abundant elements, and although less stable than water, it is stable enough to be considered as a possible biological solvent. Under one atmosphere of pressure, it is liquid at temperatures between -78 and $-33\,°C$. This 45-degree range is much narrower than the 100-degree range of water, but it covers a region of the temperature scale where water cannot function as a solvent. Henderson considered ammonia, pointing out that it is the only known substance that approaches water in its qualifications for a biological role. But in the end he rejected it on the grounds that it could not accumulate in sufficient quantity on any planet, that unlike water it does not expand near its freezing point (with the result that pools of ammonia could remain permanently frozen), and that its adoption as a solvent would give up the advantages of oxygen as a biological reagent. Henderson was vague about his reasons for thinking that ammonia could not accumulate on planetary surfaces, but he was right nonetheless. Ammonia is more easily decomposed by solar ultraviolet than is water—that is, it is split at longer, less energetic, more abundant wavelengths. The hydrogen that results diffuses into space (except on the largest planets), and nitrogen remains behind. Water is also photodecomposed in the atmosphere, but only by much shorter wavelengths than those that decompose ammonia, and the resulting oxygen (O_2) and ozone (O_3) form a very effective screen against ultraviolet radiation. Photodecomposition of atmospheric water vapor is thus self-limiting. This is not the case for ammonia.

This argument does not apply to the Jovian planets where, because hydrogen is an abundant and permanent constituent of the atmosphere, we would expect ammonia to be present. Spectroscopic observations have confirmed this expectation for Jupiter and Saturn. It is doubtful that liquid ammonia exists on these planets, but ammonia-ice clouds probably do.

Taking a broad view of the water question, we cannot prove or disprove by a priori reasoning that water is replaceable as the biological solvent. Discussions of this problem tend to be oversimplified, in that they usually consider only the physical properties of alternative solvents. They minimize or ignore the fact, which was clear to Henderson, that water is not only a solvent but is also an active participant in biochemical reactions. The elements of water are incorporated into living matter through hydrolysis and through green-plant photosynthesis (see Reaction 4-2). With a different solvent, the chemical structure of living matter would necessarily be different, as would the entire biological environment. In short, the consequences of a change of solvent would be extremely far-reaching. No one has seriously attempted to work out these consequences, nor can such an attempt reasonably be expected since it would imply nothing less than the design of a new world. And in the end, the result would be of doubtful value. The simple fact is that nobody knows whether life is possible without water, and that is unlikely to be known until an example of nonaqueous life is found.

Because water is the only substance known to function as a biological solvent, we shall take the position that it is the solvent to be expected for extraterrestrial life anywhere, barring evidence indicating the presence on the planet under consideration of a different liquid that might serve in this role.

AIRLESS WORLDS

We can thus conclude that life cannot exist on the moon or most of the other satellites of the solar system, or on Mercury, or on the asteroids because none of these objects can retain a significant atmosphere. (The asteroids are a multitude of small bodies, the largest one about 600 miles in diameter, that orbit the sun in the so-called asteroid belt between the orbits of Mars and Jupiter. The asteroid belt is the source of many of the meteorites that strike the earth.)

In the early 1960s, however, some of NASA's[2] advisers were not convinced that the moon was lifeless. Believing that "destructive alien organisms" might be hidden beneath the moon's surface, they urged the space agency to quarantine astronauts, spacecraft, and geological samples returned from space missions. Faced with contradictory advice on this question, NASA adopted the safe, if not entirely rational, line by instituting a quarantine policy designed to protect the earth against what came to be known as "back contamination." This policy was reflected in the design of the Lunar Receiving Laboratory in Houston, to which samples from the moon would be brought. It also produced a quarantine proce-

---

[2] The National Aeronautics and Space Administration. The recommendation cited here is from the report of a study conducted for NASA by the Space Science Board of the National Academy of Sciences in the summer of 1962.

dure that would isolate returning lunar astronauts for a period of three weeks as a precaution against their bringing back a nameless infection from the moon. To some observers these measures represented simple common sense; to others, they were high comedy.

As the time approached for the launch of *Apollo 11,* which was to put the first men on the moon, doubts surfaced about the need for the quarantine, especially since it would add significantly to the burden already placed on the astronauts. The public revelation that the quarantine might be relaxed set off a national debate. *The New York Times,* in an editorial on May 18, 1969, took the negative position, declaring that relaxation would have "unknowable but conceivably disastrous consequences." In reply, experts such as Edward Anders of the University of Chicago and Philip Abelson, editor of *Science,* pointed out that unsterilized material from the moon, ejected by impacting meteorites, has been arriving on the earth for billions of years; millions of tons of it have accumulated here by now. Anders offered to eat a sample of unsterilized lunar dust to prove that it is harmless. Joshua Lederberg wrote from Stanford University that had any responsible adviser believed there was a risk, NASA would have been told to cancel the manned lunar program. In the end, NASA adhered to the quarantine procedures for the first few Apollo missions, but abandoned them later.

The samples brought from the moon by the Apollo crews have been studied more carefully from more viewpoints by more different scientists in a more organized way, perhaps, than any materials ever investigated. All tests for living organisms have been negative. The same is true for attempts to detect microfossils in the samples. Chemical analysis showed the presence of carbon at a level of some 100 to 200 parts per million, most of it inorganic (carbides, for example). The evidence indicates that most of the carbon on the moon's surface originated in the solar wind, the name given to the streams of charged particles that emanate from the sun. Some small organic compounds were detected in the Apollo samples at trace levels (one to a few parts per million). Organic matter from meteorites would, of course, be expected on the moon, but it is impossible to say whether any of the traces actually found were meteoritic in origin or were contaminants introduced either by the rocket exhaust or by handling after return of the samples to the earth. The fact that no meteoritic organic matter has been identified with certainty suggests that organic compounds are destroyed on the lunar surface. In any case, it is safe to conclude that the moon is lifeless and has probably always been so.

Except for Titan, a satellite of Saturn, and possibly Triton, a moon of Neptune, all of the satellites in the solar system resemble our moon in that they lack a significant atmosphere. Ganymede and Callisto, two Mercury-sized satellites of Jupiter, are of interest in that their low densities (Table 4-1) argue that they contain much water. Current models suggest that both satellites may have subsurface oceans, but any water on their surfaces is in the form of rock-hard ice at temperatures below $-100°C$.

We now turn to objects whose masses (and in some cases low temperatures) allow them to retain an atmosphere.

VENUS

Venus is the planet nearest the earth, and it is also the planet most like the earth in mass, diameter, and density (Table 4-1). It has been known since the eighteenth century to have an atmosphere. An unbroken, highly reflective blanket of clouds makes the surface invisible from the earth, however, and explains why Venus is the third-brightest and, for many stargazers, the most beautiful object in the sky (see Plate 2). It was originally assumed that the clouds were water clouds, as on the earth, and Venus was therefore thought to have an abundance of water on its surface. Some scientists pictured the planet as covered by a vast, steaming swamp; others imagined a global ocean. In either case, the chance for life seemed excellent.

Spectroscopic results obtained in the 1930s showed the presence of considerable carbon dioxide in the Venusian atmosphere, but no water vapor. It was questionable whether water vapor could have been detected above the cloud deck, however, even with an ocean on the surface, and the idea of a wet Venus was not ruled out. Other proposals for the clouds ranged from inorganic dust to hydrocarbon smog. It was not until 1973 that several investigators independently found that the properties of Venus's clouds are best

Table 4-1　Planets and Major Satellites

Planet or satellite	Distance from sun (Earth = 1)	Mass (Earth = 1)	Density (water = 1)	SURFACE		Major atmospheric gases
				Temp. (°C)	Pressure (ATM)	
Mercury	0.39	0.055	5.4	167	0	
Venus	0.72	0.82	5.2	460	90	Carbon dioxide, nitrogen
Earth	1.00	1.00	5.5	15	1	Nitrogen, oxygen, water vapor, argon
Moon		0.012	3.3	0	0	
Mars	1.52	0.11	3.9	−55	0.006	Carbon dioxide, nitrogen, argon
Phobos		1.5×10^{-9}	1.9?		0	
Deimos		0.3×10^{-9}	2.1?		0	
Jupiter	5.19	318	1.3			Hydrogen, helium
Io		0.015	3.6	−145	0	
Europa		0.008	3.0	−145	0	
Ganymede		0.025	1.9	−145	0	
Callisto		0.018	1.8	−145	0	
Saturn	9.53	95	0.7			Hydrogen, helium
Titan		0.023	1.9	−180	1.6	Nitrogen, methane
Uranus	19.2	15	1.2			Hydrogen, methane
Neptune	30.1	17	1.7			Hydrogen, methane
Triton		0.006?	2?	−215	0.0001	Methane
Pluto	39.5	0.002	1?	−220	0.0001	Methane

fitted by droplets of concentrated (75 to 80 percent) sulfuric acid, and it is now generally agreed that this is what the clouds are. In the meantime, investigations carried out using the powerful new methods of radioastronomy and interplanetary spacecraft had shown that the mean surface temperature of the planet is around 450°C and that the atmosphere below the clouds consists almost entirely (96 percent) of carbon dioxide with a pressure, at the surface, of 90 atmospheres. Liquid water cannot exist at the surface temperature of Venus.

The high temperature of Venus is caused by the greenhouse effect: sunlight reaching the surface is absorbed by the ground and reradiated as heat, but because of the opacity of the atmosphere to infrared wavelengths, the heat cannot escape. Some evidence supports the belief that Venus once had an ocean that evaporated as the planet warmed up. Most of the water vapor was decomposed by solar ultraviolet, the hydrogen escaped, and the oxygen that remained oxidized the carbon and sulfur on the surface to carbon dioxide and oxides of sulfur. Presumably the same thing would happen on the earth if we were as close to the sun as Venus. This scenario also explains why it is that carbon dioxide on Venus is in the atmosphere, whereas on the earth it exists mainly in the form of carbonate rock. On our planet, carbon dioxide dissolves in the oceans, from which it is precipitated as the carbonate minerals calcite (limestone) and dolomite; on Venus, with no oceans, it remains in the atmosphere. It is estimated that if all the carbon on the earth's surface and in its crust were converted to carbon dioxide, the mass of this gas would approach that found on Venus.

Although conditions may have been more favorable in the remote past, it is clear at least that life has not been possible on Venus for a very long time.

THE GIANT PLANETS

Jupiter, Saturn, Uranus, and Neptune—often called the Jovian planets—are all much larger than the earth (see Table 4-1.) Among these giants, Jupiter and Saturn are supergiants: between them they contain over 90 percent of the mass of all the planets. The low densities of these four bodies imply that they are composed largely of gases and ices, and since hydrogen and helium cannot escape from their gravitational fields, they are expected to resemble the sun in their elemental composition (see Table 3-1) rather than the terrestrial planets. Observations of Jupiter and Saturn from the earth and from the Pioneer and Voyager spacecraft have in fact shown that both planets are composed predominantly of hydrogen and helium. Because of their great distance from the sun, Uranus and Neptune are poorly known, but hydrogen and the hydrogen-rich gas methane (CH_4) have been identified in their atmospheres by earth-based spectroscopy. The presence of helium is inferred, but it has not actually been observed, owing to its lack of accessible spectroscopic lines. Most of this discussion, therefore, deals with Jupiter and Saturn.

Much of what is known about the structure of the giant planets is based on theoretical models which, because of the planets' simple compositions, can be calculated with reasonable confidence. The models indicate that at their centers Jupiter and Saturn have rocky cores, larger than the earth, under pressures of millions of atmospheres and at temperatures of 12,000 to 25,000°C. These high temperatures satisfy the observation that both planets radiate approximately twice as much heat as they receive from the sun. Heat is transported to the planets' surfaces from their interiors. This implies that temperatures decrease with increasing distance from the core: at the top of the clouds—or visible "surface"—of each planet, the temperatures are −150 and −180°C, respectively. Surrounding the core is a thick stratum composed largely of metallic hydrogen, an electrically conducting form of hydrogen that is produced at high pressures. Above this is an envelope of molecular hydrogen mixed with helium and small quantities of other gases. Near the top of the hydrogen-helium envelope are layers of clouds, the compositions of which are determined by local temperatures and pressures. Clouds of water ice, perhaps containing droplets of liquid water, form where the temperature approximates 0°C. Somewhat higher are clouds of ammonium hydrosulfide, and above these, at temperatures in the neighborhood of −115°C, clouds of ammonia ice are formed.

This description assumes a solar composition for the two planets, with hydrogen making up 90 percent or more of the atmosphere by volume or number of molecules. In such atmospheres we expect carbon, oxygen, and nitrogen to be present almost entirely as methane, water, and ammonia, respectively. These gases, along with hydrogen, have been detected on Jupiter, all except water in amounts expected for an atmosphere of solar composition. The spectra are deficient in water, possibly because water vapor condenses out at relatively deep levels. Besides these gases, carbon monoxide and traces of the simple organic molecules ethane (C_2H_6), acetylene (C_2H_2), and hydrogen cyanide have all been reported for the Jovian atmosphere. The source of the bright colors of Jupiter's clouds—reds, yellows, blues, browns—has not been positively identified, but both theoretical and laboratory studies suggest that sulfur, sulfur compounds, and perhaps red phosphorus are responsible.

The discovery of water vapor and simple organic compounds in the upper atmosphere of Jupiter and the possibility that clouds of liquid water exist at lower levels raise the question of chemical evolution on the planet. At first glance, it would appear that complex organic compounds like those formed in laboratory simulations of the primitive earth (see Chapter 3)—and possibly even indigenous life—are to be expected as a matter of course in the reducing atmosphere of Jupiter. Indeed, even before water vapor and organic molecules had been detected in the Jovian atmosphere, Carl Sagan had expressed the view that "of all the planets in the solar system Jupiter is the one with the largest a priori biological interest." Actual Jovian conditions, however, have disappointed these hopes.

The atmosphere of Jupiter does not favor the production of complex organic compounds for a number of reasons. In the first place, at the high temperatures and pressures that exist in the bulk of this very strongly reducing environment, hydrogen destroys organic molecules by converting them into methane, ammonia, and water. Moderately reducing—that is partly oxidized—gas mixtures are more favorable to significant organic synthesis than strongly reducing ones, as Urey pointed out long ago. For example, the synthesis of glycine, the simplest amino acid, cannot proceed spontaneously in the mixture of gases—methane, ammonia, and water—available in the Jovian atmosphere. Rather, it requires free energy, as shown in Equation 4-4 below. On the other hand, the synthesis can occur without input of energy in a less strongly reducing mixture containing carbon monoxide, ammonia, and hydrogen, as in Equation 4-5:

$$2CH_4 + 2H_2O + NH_3 + 48,800 \text{ calories} \longrightarrow C_2H_5O_2N + 5H_2 \quad (4\text{-}4)$$

$$2CO + NH_3 + H_2 \longrightarrow C_2H_5O_2N + 19,000 \text{ calories} \quad (4\text{-}5)$$

In the presence of free hydrogen, as on the Jovian planets, equation 4-4 can be read from right to left to show that glycine reverts spontaneously to methane, water, and ammonia. Experiments with realistic gas mixtures to learn how much organic synthesis may be possible on Jupiter have not yet been reported. Such experiments are difficult to perform because of the large proportions of hydrogen and helium required (some published "Jovian" organic synthesis experiments contain no hydrogen at all), but to use anything less produces results of doubtful value.

Jupiter and the other large planets also lack suitable surfaces on which organic products that may form in the atmosphere can accumulate and interact, an important factor in considering possible sites of chemical evolution. Any such evolution must therefore take place in the atmosphere, presumably in the water clouds. But the atmosphere of Jupiter is not a stable environment like the oceans of the earth. It is more like a gigantic furnace in which vertical currents constantly move hot gases from the interior of the planet to the top of the atmosphere, where they radiate their heat to space, while cold gas is simultaneously transported downward to be reheated. The visible turbulence in the clouds of Jupiter is a sign of this strong convection (see Plate 3). How much chemical evolution can occur in this environment before organic molecules formed by sunlight in the high atmosphere are transported to deep, hot regions where they are destroyed? The answer appears to be: Very little. Calculations indicate that a parcel of gas at the level of the water clouds is transported to a level in the atmosphere where the temperature exceeds 200°C in a matter of days. Organic compounds will therefore be destroyed in a short time and their carbon, nitrogen, and oxygen will be reconverted into methane, ammonia, and water.

Even allowing for uncertainties in the calculations, it is clear that conditions in the atmosphere of Jupiter are not favorable for chemical evolution. Furthermore, Jupiter is not only a furnace but, as we saw, a reaction vessel as well, and this rules out any possibility that organic molecules might be stabilized against heat by the high pressure of the deep atmosphere. Thus we have to conclude that the lifetime of organic structures on Jupiter is too short for any extensive organic synthesis to have occurred there. The same considerations hold for Saturn (see Plate 4), and they probably hold for Neptune as well. Uranus is a mystery at the time of this writing, but it is a safe guess that it will prove to be no more hospitable than the other giant planets.

TITAN, TRITON, AND PLUTO

Titan, the largest moon of Saturn, is the only satellite known to have a dense atmosphere. Much speculation about this unusual Mercury-sized body came to an end in 1980, when *Voyager I* approached to within 5000 kilometers of Titan's surface and sent back a large amount of data on Titanian chemistry and physics. (A summary of these data and of the work of many investigators appears in the papers by Stone and Miner and by Pollack listed in the Bibliography.)

As Table 4-1 shows, the atmospheric pressure at the surface of Titan is 1.6 atmospheres. Its major gases are nitrogen (90 percent or more) and methane (1 to 10 percent), and small quantities of ethane, acetylene, ethylene (C_2H_4), and hydrogen cyanide have also been identified. These latter are photochemical products, some of which we have also seen in the atmosphere of Jupiter, formed by solar ultraviolet light acting on methane and, in the case of HCN, nitrogen gas. Ammonia would be frozen solid at the low temperature of Titan ($-180°C$). Carbon monoxide and carbon dioxide molecules have also been detected in the Titanian atmosphere. These molecules were not expected, since it is assumed that Titan's oxygen is frozen out in the form of water ice on its surface. A possible source of oxygen is the water contained in infalling meteorites. (This may also be the source of oxygen for the carbon monoxide seen in the atmosphere of Jupiter.)

The surface of Titan is obscured by an atmospheric haze, which is believed to be composed of large hydrocarbon molecules produced photochemically from methane—a kind of smog (see Plate 5). Growth of the smog particles by cohesion may result in grains large enough to fall to the surface and form drifts. In addition, at the low temperature of Titan there is the possibility of liquid ethane, which, it has been suggested, may form an ocean. Titan may thus have an abundance of both organic matter and solvent. Yet it is an unlikely place for life or for chemical evolution, owing to its low temperature, which is close to that of liquid air. At $-180°C$, the rates of reactions in solution are too low for much to have happened in the way of chemical evolution, even over the lifetime of the solar system. The chemistry that occurs in the atmosphere has its source of energy in solar ultraviolet photons. But the chemistry of solutions depends on thermal energy, which Titan has little of. The organic chemistry of Titan should nevertheless make a fascinating study for some future space probe.

Triton, the largest satellite of Neptune, is difficult to observe and is poorly known. It was recently found to possess a tenuous atmosphere of methane, but in view of its size and low temperature, it is possible that it actually has a larger atmosphere. Its surface temperature is lower than Titan's, well below the freezing point of liquid air.

Pluto is the smallest and also the most remote of the planets. Its orbit is as far from Neptune's, on average, as Saturn's is from the sun. Its very small mass and its unusual orbit indicate that Pluto did not originate in the same way as the other planets. It has been suggested that it was originally a satellite of Neptune and that it should be considered as an asteroid, not as a major planet. Be that as it may, it has been found to have a slight atmosphere of methane and to have solid methane on its surface. The surface temperature of Pluto is even lower than that of Triton. A less likely place for life would be hard to imagine.

Having considered all of the planets except Mars (and the earth) as possible biological habitats, we conclude that none of these now provides a suitable environment for life, although in some cases it is possible that conditions were more favorable in an earlier epoch. It goes without saying that much remains to be learned about the solar system, but it is unlikely that future findings will alter this conclusion. All the essential points of the argument presented here were known or surmised before the launch of two Viking spacecraft to Mars in 1975. By that time, only Mars appeared to be a possible abode of extraterrestrial life. In the next chapter, we take up the strange history of the Martian investigations that culminated in the Viking mission.

CHAPTER FIVE

Mars: Myth and Reality

Our knowledge of Mars steadily progresses. Each opposition adds something to what we knew before. Since the theory of life on the planet was first enunciated some fifty years ago, every new fact discovered has been found to be accordant with it. Not a single thing has been detected which it does not explain. Every year adds to the number of those who have seen the evidence for themselves. Thus theory and observations coincide.

E. C. Slipher, *The Photographic Story of Mars* (1962)

In *The Golden Bough,* anthropologist Sir James Frazer told us that Mars was originally a god of vegetation, not war. Roman farmers prayed to him for the success of their crops, and the vernal month of March was consecrated to him. In view of this ancient association of the god Mars with the awakening of plant life in the spring, it seems fitting that the planet Mars should be the least hostile of all the extraterrestrial bodies in the solar system, the planet other than the earth that comes closest to providing a suitable habitat for life.

Although only about half the diameter of the earth, Mars looks remarkably earthlike from a distance, and it does share certain similarities. In one of the earliest telescopic studies of the planet in 1659, Christian Huygens (whose views on extraterrestrial life were quoted on page 55) discovered a permanent marking on the Martian surface that enabled him to measure the planet's rotational period. He found that Mars, like the earth, rotates on its axis once in 24 hours. Later, more accurate measurements showed that the length of the Martian solar day is exactly 24 hours, 37 minutes, and 22 seconds—a period that became known as a "sol" during the Viking mission in order to avoid confusing it with the terrestrial day. Furthermore, in the present era the spin axis of Mars tilts at an angle of 25 degrees from the plane of its orbit, compared to the 23.5 degree tilt of the earth.[1] This means that Mars has seasons like the earth's, as first one hemisphere and then the other leans toward the sun. The Martian year is 687 earth days in length (669 sols), or about six weeks short of two earth years, so that the Martian seasons last roughly twice as long as those on the earth. Because of the eccentricity of Mars' orbit, however, Martian seasons are not of nearly equal durations as our planet's are. On Mars, for example, northern summer (and southern winter) lasts 178 sols, while northern winter (and southern summer) lasts 154 sols. On the earth, these seasons last 94 and 89 days, respectively.

The impression of an earthlike planet is reinforced by seasonal changes on the Martian surface that can be seen through the telescope. The most striking of these changes is the annual advance and retreat of the polar ice caps (see Plate 6). Other, more subtle alterations occur at lower latitudes, where the Martian surface is broken up into bright and dark areas (see Plate 7). The bright areas, formerly called deserts, are reddish orange in color. The dark areas, in the past called maria (seas) because they were thought to be bodies of water, have been variously described as gray, brown, blue, and green. Seasonal changes in color and contrast—the maria appeared a dark bluish green in the late spring

---

[1] The tilt of Mars' axis varies between 15 and 35 degrees over time, largely because of the gravitational pull of Jupiter. That of the earth changes by plus-or-minus only 1 degree from its present value.

and summer, but faded to a brownish tone in fall and winter, then darkened again in the spring—were reported by astronomers in the nineteenth century. Such changes made it appear more likely that vegetation rather than water covered these areas, an idea that was first advanced in 1860. Some observers also described a network of pencil-thin, straight lines extending for hundreds of miles over the Martian surface. These lines, called *canali* (channels) by Giovanni Schiaparelli (1835–1910), the foremost Mars-mapper of the nineteenth century, changed seasonally like the maria—they were dark in the local spring and summertime, but faded in fall and winter. Schiaparelli noted that the canali looked like the work of intelligent beings, but he did not commit himself to this interpretation.

These intriguing observations, which were made possible by improvements in telescopes in the nineteenth century, convinced some people that direct evidence for life on another planet had at last been achieved. One of those stirred by the new findings was Percival Lowell (1855–1916), an American who would found the Lowell Observatory in Flagstaff, Arizona, for the express purpose of studying Mars. Lowell's place in the extraordinary history of Martian biological investigations is special enough to require a more detailed examination.

THE LEGACY OF PERCIVAL LOWELL

Percival Lowell belonged to an eminent New England family: Abbott Lawrence, his brother, became president of Harvard University, and his sister Amy was the Imagist poet. Lowell was not trained as an astronomer—he devoted himself to studies of Japanese and Korean culture, subjects on which he wrote a number of books—and his fascination with Mars developed relatively late in life. According to William Graves Hoyt, author of a recent biography of Lowell, astronomy had long been among Lowell's many enthusiasms, and he was inspired by Schiaparelli's discovery of canali on Mars. Schiaparelli's description, it appears, all but convinced Lowell that Mars was inhabited by intelligent beings. This belief, or near-belief, led him to commit his considerable wealth and talent to the founding of the Flagstaff observatory. Dedicated to the study of Mars, the observatory opened in May, 1894. By July of the same year, barely two months after operations began, Lowell was ready to formulate his very definite views about life on Mars— views from which, Hoyt notes, he did not deviate for the rest of his life.

Although Lowell came to Martian studies late in his career, his authority and influence in matters relating to Mars soon became quite important. His observatory was well equipped and staffed, and it occupied a superior observing site, a fact that Lowell did not neglect to mention when others failed to confirm his observations. Furthermore, the observatory spent almost its entire effort on Mars and missed no opportunity to study it. Lowell therefore accumulated a vast amount of systematic information on the planet, and these intensive studies made him probably the best-informed Mars-watcher of his day. (Criticisms directed at Lowell during his lifetime and later aimed less at his data than at his interpretations of them.) Finally, Lowell tirelessly communicated his great enthusiasm for his subject and his unwavering confidence in the correctness of his conclusions to the general public in books, articles, and public lectures. Through these and other efforts, he succeeded in making Mars a topic of general interest, not one reserved for specialists alone.

Lowell's theory was simple enough. It began with the premise that the polar caps of Mars are composed of water ice. To support this belief, he pointed to a dark blue band, or collar, that appeared around the caps as they began to shrink in the spring and that dwindled with them. Only liquid water, produced by melting of the caps, could explain the collar, he argued, and he often referred to these bands as "polar seas." Lowell knew that, aside from the polar regions, Mars is very dry. Its dark areas could not be bodies of water because, although they changed color seasonally as if drying up, the water apparently lost from them did not turn up anywhere else on the planet. And, as others had pointed out, if the maria contained water, they would reflect sunlight, yet such reflections were never seen. Given the dryness of the planet, Lowell deduced that the seasonal disappearance of one polar cap, coupled with growth of the other, meant that water was transferred from one pole to the other: "Meteorological conditions carry [water] to deposit at one pole, then liberate it and convey it to imprisonment at the other, and this pendulumlike swing of water is all in the way of moisture that the planet knows." This semiannual transfer of water was accompanied by a darkening of the dark areas that spread wavelike "across the face of the planet from one pole to the other in the course of a Martian six months." Such regular darkening, he was convinced, demonstrated that plant life exists on Mars. The observations proved, he wrote, that the condition of the planet "is not only compatible with life, but that vegetal life shows itself there as patently as could possibly be expected, and that nothing but vegetation could produce the observed phenomena."

In Volume 1 of the *Annals of the Lowell Observatory,* Lowell continues:

> Now if there were any life of an order higher than the vegetal upon the planet—an order capable of something more advanced than simply vegetating, an order able to turn natural conditions to its own ends—its first and final endeavors would be to contrive means to use every particle of that necessary and yet scanty sustainer of life, water. For there is no organism which can exist without water. In short, irrigation for agricultural purposes would be the fundamental Martian concern. . . .

Lowell then summarizes his observations on the canali and concludes:

> These are just the features a gigantic system of irrigation would present. Upon the above results of the observations is based the deduction I have here put forward,—(1) of the general habitability of the planet; (2) of its actual habitation at the present moment by some form of local intelligence.

And so Lowell arrived at the belief in a civilization on Mars. Abetted by his French contemporary, the astronomer Camille Flammarion (1842–1925), he popularized the familiar drama of a courageous race of Martians, superior to ourselves, struggling for survival on a desiccated and dying planet. Lowell saw nothing spec-

Plate 1   The Great Nebula in Orion. This enormous mass of
gas and dust surrounding the middle star of Orion's sword is
another illustration of the abundance of hydrogen in the
universe. The radiations from several hot stars in the nebula
excite its gases and cause them to radiate at their characteristic
frequencies. The red light in the photograph comes from
hydrogen, the blue from oxygen and nitrogen, and the white
from a mixture of gases (copyright 1959, California Institute
of Technology).

Plate 2 Venus, photographed by *Mariner 10* in the ultraviolet to bring out cloud structure. The blue color was added (NASA/JPL).

Plate 3 The Great Red Spot, a long-lived feature in the atmosphere of Jupiter, surrounded by turbulent clouds (*Voyager* photograph, NASA/JPL).

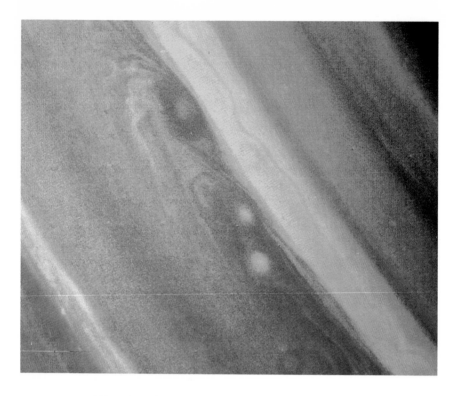

Plate 4 The northern hemisphere of Saturn, photographed through color filters, showing atmospheric storms (*Voyager* photo, NASA/JPL).

Plate 5 Photochemical haze layers in the atmosphere of Titan, photographed in false color by *Voyager* (NASA/JPL).

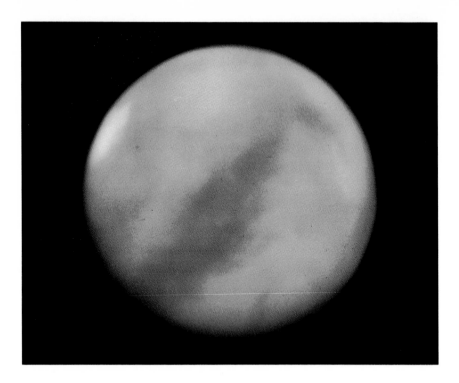

Plate 6 Taken at Mt. Wilson in 1954, this is one of the best earth-based photographs ever made of Mars. Bright and dark regions of the planet are visible; the shrinking south-polar ice cap appears in the upper left due to inversion by the telescope. Now compare this view with that in Plate 7 (copyright 1965, California Institute of Technology).

Plate 7 This photograph of the sunlit hemisphere of Mars
was taken by *Viking 1* as the spacecraft approached the planet
in 1976. North is at the top. Frost, haze, and much surface
detail can be distinguished. The large canyon, Valles Marin-
eris, is toward the top, and the Argyre basin is toward the
bottom. The ochre color is due to iron oxide (NASA/JPL).

Plate 8   The Don Juan Pond in Antarctica (photo by Roy Cameron).

Figure 5-1. Lowell's 1907 globe of Mars, with canals. The place names were given by Schiaparelli (Lowell Observatory Photograph).

ulative in these ideas. In a typical passage from his book, *Mars as the Abode of Life,* he writes:

> In our exposition of what we have gleaned about Mars, we have been careful to indulge in no speculation. The laws of physics and the present knowledge of geology and biology, affected by what astronomy has to say of the former subject, have conducted us, starting from the observations, to the recognition of other intelligent life.

The idea of a Martian civilization, although embraced enthusiastically by the public, was viewed with scepticism by scientists even in Lowell's day, and it did not survive his death. The "canals"—always controversial and now known not to exist—were probably an illusion caused by viewing difficulties. The rest of Lowell's theory, however—the polar ice, the movement of water, the vegetation—not only survived him, but took on new life. It was as if, relieved of its burden of heroic Martians, the Lowellian thesis of an earthlike Mars had acquired scientific respectability and could be accepted as a reasonable approximation to the truth. Lowell's views would eventually be disproved in every significant detail. Yet —and this is the strange part of the story—as observations of Mars continued, they seemed increasingly to show that Lowell was right. As a result, the Lowellian view was widely accepted for most of the Twentieth century.

The epigraph at the beginning of this chapter, quoted from a book by a longtime associate of Lowell's, summarizes the state of affairs as they appeared to one observer in 1962. The optimism that shines through Slipher's statement could, in fact, be defended in 1962. Unfortunately for this view, new results soon to be obtained would show that this optimism was unfounded and that the Lowellian picture of Mars, with or without canal-building Martians, was pure fantasy. Within a few years, scientific ideas about the planet would take an entirely different turn. The rise and fall of Lowellian Mars in our time is our subject in the rest of this chapter.

PRE-1963, OR LOWELLIAN, MARS

Ice Caps

The waxing and waning of the Martian ice caps proved to early observers that Mars has an atmosphere of some kind, but its composition and its quantity remained a mystery for a long time. Carbon dioxide, now known to be the major component of the Martian atmosphere, was first identified on Mars by the well-known Dutch-American astronomer, Gerard P. Kuiper (1905–1973) in 1947. To make this finding, Kuiper used infrared spectroscopy. In planetary spectroscopy generally, sunlight re-

Figure 5-2   This photograph, taken in 1911, shows why it is so difficult to study Mars from the earth. Here Mars emerges from behind the moon, following an occultation (eclipse) by the latter. Although it was relatively close to the earth at the time, the planet appears no larger than a small lunar crater (Lowell Observatory Photograph).

flected from a planet is collected by a telescope and broken up by a prism or grating into its spectrum—in this case, its infrared spectrum. This spectrum is then compared to a similarly prepared spectrum from the moon, for example, or, depending on the question being investigated, from a different area of the same planet. Different substances absorb light of different wavelengths, a fact that makes chemical identifications possible. By comparing the spectrum from a planet that has an atmosphere with one obtained from the moon, which has no atmosphere, once the absorption by the earth's atmosphere is subtracted out, what remains is the spectrum of the planet's atmosphere. Because the amount of energy absorbed depends on the amount of absorbing material, such spectra contain quantitative as well as qualitative information. Thus, the observer can say not only what the absorbing gas in the light path is but how much of it there is.

The wavelength region that lies beyond the red end of the visible spectrum, called the infrared, is very rich in specific chemical absorptions, When Kuiper compared the infrared reflections from Mars with those from the moon, he found a diminution in the energy at several wavelengths in the neighborhood of 1.6 micrometers[2] in the Martian spectrum; these wavelengths corresponded to known carbon dioxide absorptions. Kuiper estimated that the amount of CO_2 over a given area of Martian surface is twice that over the same area on the earth. From this, he calculated the pressure of CO_2 on Mars, taking into account the lower force of gravity on Mars compared with that on the earth. He arrived at a pressure equal to that of 0.26 millimeter of mercury, or 0.35 millibar.[3] Here Kuiper erred. His estimate was too low by a factor of about 16, an error that had important consequences because it allowed Kuiper to argue that the Martian polar caps could not be composed of frozen carbon dioxide (dry ice). If the carbon dioxide pressure were as low as his calculations indicated, an unrealistically low temperature would be required to freeze the gas out of the atmosphere. It was discovered some years later that Kuiper had incorrectly calculated the CO_2 pressure, but this finding had no effect on the course of events.

The only other substance that was reasonable for the caps was

---

[2] A micron, or micrometer, is one one-millionth of a meter.

[3] A millibar (abbreviated mb) is a pressure of 100 newtons per square meter. The pressure of the earth's atmosphere is 1013 mb, or 1.013 bars. A 760-millimeter column of mercury at the earth's surface exerts a pressure of 1013 mb.

frozen water—ice, snow, or frost—but other astronomers had already searched without success for water vapor in the Martian atmosphere. Kuiper therefore proceeded to examine the north polar cap of Mars directly by infrared spectroscopy. The analysis was difficult, owing to the small size of the ice cap, but by modifying his spectrometer to improve its sensitivity and by repeating the observations many times, Kuiper finally convinced himself that "the Mars polar caps are not composed of CO_2 and are almost certainly composed of H_2O frost at low temperature." The note of caution discernible in the second part of this statement reflects the fact that the spectrum of the polar cap did not quite match the spectrum Kuiper had obtained for terrestrial snow.

Here Kuiper erred again—the seasonally variable portions of the caps are actually frozen carbon dioxide, not water—but his error would not be discovered for nearly 20 years. Instead, the incorrect result was seemingly confirmed by another investigator, Audouin Dollfus of the Paris Observatory, using a different method, one based on the polarization of reflected light. Ordinary, unpolarized sunlight consists of electromagnetic waves that vibrate in all directions in the plane perpendicular to the direction of propagation of the light ray. In light that has been reflected or scattered, however, or that has passed through certain kinds of materials (such as polaroid), the waves vibrate in only one direction. Such light is said to be polarized. The degree of polarization of reflected light depends on the angle of vision, as well as on the structure, transparency, and other physical properties of the reflecting surface. Dollfus, who had much experience with planetary polarization measurements, decided to apply the method to the ice-cap problem.

As had Kuiper before him, Dollfus noted that the caps were small and difficult to study. He succeeded in making some measurements, however, and found that the polarization of the caps was much less than that of ice, frost, and snow on terrestrial mountains observed at the same angle. He then performed laboratory experiments showing that the polarization of frost deposits could be made to resemble that of the Martian ice caps if, first, the frost were deposited on a cold surface with a low atmospheric pressure (as would be the case on Mars) and, second, if the deposits were

partly sublimed—that is, evaporated without melting—by exposing them to an arc lamp, as might happen to Martian ice caps exposed to the sun. He concluded that the ice caps were probably
deposits of hoarfrost.

Dollfus did not perform comparable experiments with solid
CO_2, but his apparent confirmation of Kuiper's result convinced
many students of Mars that the ice-cap question was solved. The
following quotation, for example, reflects the views of a panel of
informed scientists, many of whom later made important contri-
butions to our knowledge of Mars, that was appointed by the Space
Science Board to advise NASA in the early stages of its planetary
program:

> Infrared reflection spectra of the polar caps show con-
> clusively that they are not composed of frozen carbon
> dioxide, the only condensible substance which might
> be expected besides water; the reflection spectra are
> also consistent with the assumption that the polar caps
> are made of ice. . . . The polarization data indicate
> that the polar caps are composed of hoar frost. . . .

Later in its report, this same panel drew an inference identical to
the one reached by Lowell in 1898, 63 years before. In their view,

> Since the polar caps are composed of frozen water,
> their seasonal retreat directly suggests that there is
> some water vapor in the Martian atmosphere. Because
> of the alternating formation of polar caps in opposite
> hemispheres, the circulation in the lower atmosphere
> must be such that water vapor is transferred from one
> hemisphere to the other.

Atmospheric Pressure

Mutually supporting errors also produced a spurious conclusion
about the atmospheric pressure, another crucial biological parame-
ter. And once again, the effect was to make Mars appear more
earthlike than it actually is. In the Lowellian period, the two principal methods used to estimate the Martian atmospheric pressure

were photometry and polarimetry. Light is scattered by gas molecules. It is this scattering that accounts, for example, for the blue sky: the atmosphere scatters incoming sunlight in all directions, but because light of short wavelengths (blue and violet) is scattered very much more than long-wavelength (red) light, we see a blue sky when we look away from the sun. Since scattering by its atmosphere affects the surface brightness of a planet, measurement of that brightness at different wavelengths and under different thicknesses of atmosphere (obtained by viewing the planet at different angles) should provide a way to estimate atmospheric pressure. Furthermore, since scattered light is polarized, polarization measurements should provide a check on the result.

But scattering depends not only on the wavelength of the light and the mass of the atmosphere but also on the atmospheric composition and the presence or absence in it of dust and other suspended particles, and therein lies the snag. In order to allow for these and other complications, such as polarization by the planetary surface, investigators before 1963 had to make some unverifiable assumptions before they could derive the atmospheric pressure from their data. The result, in the words of Claude Michaux and Ray Newburn of the Jet Propulsion Laboratory (JPL), was that "Each successive worker pointed out the 'unwarrantable assumptions' of his predecessors and proceeded to make a new set of his own."

Despite the difficulties, a dozen or so attempts were made after Lowell's time to apply photometric and polarimetric methods to the surface-pressure problem. Their generally concordant results were reviewed by the French astronomer Gerard De Vaucouleurs in an influential book on Mars, the English edition of which appeared in 1954. De Vaucouleurs concluded that the most probable value for the Martian surface pressure was 85 ± 4 millibars. (This was the perfect Lowellian result. In *Mars as the Abode of Life*, published in 1908, Lowell had applied photometric arguments to the problem and arrived at a pressure of 64 millimeters of mercury, or 85 millibars!) After reexamining the evidence, the panel of experts quoted above gave its judgment: "It is unlikely that the true surface pressure differs by as much as a factor of 2 from 85 millibars."

Actually, the true surface pressure differs from 85 millibars by a factor of more than 10.

Vegetation

Lowell based his belief that the dark areas of Mars were covered with vegetation on their blue-green color and on their observed seasonal color changes. In the springtime, what he called a "vernal progression" or "verdure wave" moved through these areas and along the canals, starting near the edge of the dark collar surrounding the shrinking ice cap and proceeding toward and beyond the equator. He estimated the speed of the wave at 51 miles per day. In Lowell's scheme, the wave of deepening color proved the sprouting of vegetation as water became available at lower latitudes in its regular swing through the atmosphere from one pole to the other. Lowell recognized that the direction of this wave of darkening, as it came to be called, was opposite to that seen on the earth, where the springtime growth of plants starts in temperate latitudes and moves poleward. But, he argued, this was just what one would expect on a planet where life is limited by the availability of water.

Telescopic observations made after Lowell's time confirmed the existence of the dark polar collar and of seasonal changes in the maria. These phenomena are now thought to result from redistribution of dust by seasonal winds. Or perhaps the polar collar is simply an optical effect produced by a glazed layer of frozen carbon dioxide exposed by sublimation of the overlying carbon dioxide frost. For decades following Lowell's death, however, the vegetation scenario reigned, and, by 1960, it seemed close to being proven.

The story began modestly in 1947–1948 when, following the conclusion that the Martian polar cap was composed of water ice, G. P. Kuiper turned his attention to what he called the "green areas" of Mars. His plan was to compare the light reflected from these areas with the spectra obtained from higher plants, lichens, and mosses. Lichens are symbiotic associations between a fungus and an alga. They are green or greenish in color and, like higher plants, they carry out photosynthesis by means of chlorophyll. Extremely hardy, these organisms inhabit cold, dry, forbidding places where few other living things survive.

Kuiper found no resemblance between the higher plant and lichen spectra. Whereas the plant spectrum showed a dozen or so peaks and valleys in its visible and infrared regions, the lichen spectrum appeared featureless and nearly flat over the same spectral range. Dry mosses produced a spectrum like that of lichens. Kuiper did not, for technical reasons, obtain a complete spectrum of the green regions of Mars. Instead, he measured their reflections at four different wavelengths. These observations convinced him that the Martian spectrum was unlike that of green plants and similar to that of lichens and mosses. A featureless spectrum hardly constituted strong evidence for the existence of either of these life forms on Mars, and a nonbiological explanation for the wave of darkening had been proposed. According to this hypothesis, the seasonal changes occurred when inorganic materials on the Martian surface absorbed water vapor from the atmosphere as it moved across the planet in the spring, then lost it in the dry fall atmosphere. Many such absorptive substances are known, and some of them do change color with the uptake and loss of moisture. Ernst Öpik, a noted Estonian-British astronomer, disposed of this hypothesis in 1950 with the argument that dust storms, recognized telescopically as yellow clouds that sometimes obscure the entire planet, would have covered the dark areas long ago if they had been simply inorganic deposits. The fact that the same areas always reappear after such storms, Öpik said, shows that they have regenerative powers.

Considering all the evidence and giving due weight to Öpik's argument, Kuiper concluded that the case for living things in the dark areas "appears very good." He thought it improbable, however, that Martian lichens were identical to those on the earth, because this would imply a highly unlikely parallel evolution—and besides, our lichens do not change color in the fall.

Kuiper's case was, at best, only suggestive, but it was soon strengthened by a spectacular result obtained by W. M. Sinton, a young American astronomer. As Kuiper had done, Sinton investigated the light reflected from Mars, but instead of scanning the whole spectrum, he concentrated on a narrow region in the infrared—in the neighborhood of 3.5 micrometers—where carbon-hydrogen bonds absorb strongly. Since all organic matter contains such bonds, Sinton argued that if plant life were responsible for the wave of darkening, that fact should be revealed by absorption in this region. Tests performed on lichens, mosses, and dried leaves confirmed that the light they reflect does show this absorption feature. Sinton then examined Mars over a period of four nights and found an absorption band centered at 3.46 micrometers, exactly where the tested plant material had absorbed. Two years later, in 1958, he repeated the observations, but with better equipment and with the 200-inch telescope of the Palomar Observatory. This time, he was able to analyze the light from the dark and bright regions separately. He found three bands near 3.5 micrometers—all attibuted to organic compounds—in the light that came from the dark regions. The absorptions were weak or absent in light from the bright regions. On the face of it, a stronger confirmation of Lowell and Kuiper could hardly have been imagined.

The Space Science Board panel was not convinced by the Sinton absorption bands, stating that "the possibility that they arise from a combination of inorganic substances does not seem to have been explored sufficiently." On the question of Martian life, however, it concluded:

> The evidence taken as a whole is suggestive of life on Mars. In particular, the response to the availability of water vapor is just what is to be expected on a planet which is now relatively arid, but which once probably had much more surface water. The limited evidence we have is directly relevant only to the presence of microorganisms; there are no valid data for or against the existence of larger organisms and motile animals.

THE REAL MARS

Atmospheric Pressure

The unveiling of the real Mars, which we now take up, illustrates a profound truth stated many years ago by two philosophers of science, Morris R. Cohen and Ernest Nagel: "On the whole it may be said that the safety of science depends on there being men who care more for the justice of their methods than for any results obtained by their use."

The "delowellization" of Mars began with a single spectrogram of exceptional quality obtained at the Mt. Wilson Observatory in April, 1963, and analyzed by Lewis Kaplan, Guido Münch, and Hyron Spinrad of JPL and the California Institute of Technology (Caltech). The spectrogram showed infrared absorption lines of carbon dioxide and, for the first time, water vapor in the Martian atmosphere. The CO_2 spectrum was of particular interest because it showed both weak absorption lines, whose width depended on the amount of CO_2 but not on the total atmospheric pressure, and strong lines whose width depended on both. It was thus possible for the authors to estimate the abundance of CO_2 on Mars *and* the total Martian surface pressure. Most important, the calculation of pressure could now be made from known physical principles without recourse to the unverifiable assumptions that had vitiated all earlier estimates.

The analysis by Kaplan, Münch, and Spinrad gave an unexpected result: the pressure of the Martian atmosphere appeared to be much lower and its CO_2 content much higher than they were thought to be at the time. Specifically, the best estimates of Kaplan and his coworkers were 25 millibars for the total atmospheric pressure and 4 millibars for the pressure of CO_2, compared with the accepted values of 85 millibars and 2 millibars, respectively. The authors attached fairly large errors to their estimates, owing to uncertainties in some of the measurements (all made from the one photographic plate), and they pointed out that further observations could be expected to yield improved estimates. It was eventually shown that even 25 millibars was much too high a value for the Martian surface pressure.

The 1964 paper by Kaplan, Münch, and Spinrad marks the opening of the post-Lowellian era of Mars science. Considerable effort now went into redetermining the Martian atmospheric pressure and composition, not only because these numbers had inherent interest but also because accurate knowledge was needed for designing spacecraft that would land on the planet. When Mars next approached the earth in 1965, the Martian atmosphere came under close scrutiny from earth-based observatories and from *Mariner 4,* the first U.S. spacecraft launched toward Mars.

The next surprise in the unfolding Martian drama came with the accurate and enlightening data returned by *Mariner 4*. The method employed on *Mariner 4* to measure atmospheric pressure was new for Mars investigations. It began with the trajectory of the spacecraft, which would take it behind the planet in such a way that it would be eclipsed, or occulted, by Mars for nearly an hour. For a period of about two minutes preceding the actual occultation, *Mariner 4*'s radio beam was refracted, or bent, as it passed through the Martian atmosphere on its way back to the earth. The same thing happened when the spacecraft emerged from behind Mars 54 minutes later. The amount of refraction was accurately measured when the radio signal arrived back at the earth, and, because refraction is a function of atmospheric density, a profile emerged of the pressure from the top of the Martian atmosphere to the point on the surface where the spacecraft entered or exited occultation.

The surface pressures thus obtained were surprisingly low—4 to 7 millibars, depending on the temperature of the atmosphere and its actual carbon dioxide content (which was not precisely known at the time). Pressures such as these would be found in the earth's atmosphere at altitudes of about 20 miles. It seemed at first that such low pressures must refer to high points on the surface of Mars and not the planet as a whole. This, however, did not turn out to be the case. Many measurements of the Martian pressure have been made since 1965, by various methods and from a variety of observation points. These range from Mars-wide averages obtained spectroscopically from the earth to local measurements made on the Martian ground by pressure sensors aboard the Viking landers. All agree in showing that the average pressure—which varies somewhat with the Martian season and locale—is well under 10 millibars. Estimates made by different authors range from 5 to 7 millibars, and we can adopt 6 millibars as a reasonable value. The pressure in the Hellas basin, one of the lowest points on Mars, would then be about 8.6 millibars, and that on the top of Olympus Mons, the highest point, about 0.5 millibar.

Composition of the Atmosphere and Ice Caps

The *Mariner 4* results made it quite clear that carbon dioxide, whose pressure on Mars had been estimated at 4 millibars by Kaplan, Münch, and Spinrad, must be a major component of the Martian atmosphere, not a minor one as was believed in the days of the 85-millibar atmosphere. (The Viking mission later established that carbon dioxide makes up 95 percent of the Martian atmosphere.) Other gases detected on Mars before Viking were water vapor—this had appeared on the plate analyzed by Kaplan and his coworkers—and small amounts of photolytic products formed from it and from carbon dioxide by the action of sunlight—to wit, oxygen, ozone, atomic hydrogen, and carbon monoxide. The abundance of water vapor was estimated at 14 micrometers precipitable, meaning that if all the water vapor in the atmosphere were condensed to liquid, it would make a layer 14 micrometers thick. This much water vapor would exert a surface pressure on Mars equal to $1/8000$ that of the atmospheric carbon dioxide, or 0.5 microbar.[4] The pressure of water vapor on the earth's surface is, on average, 10,000 times greater than this. Such a disparity has important biological implications, and these are treated more fully in later chapters.

The importance of carbon dioxide in the Martian atmosphere led Robert Leighton and Bruce Murray of Caltech to reexamine the question of the composition of the polar caps. In 1966, Leighton and Murray published a theoretical study of the Martian heat budget that enabled them to predict the temperature at any Martian latitude at any time of the year. Mars is expected to be colder than the earth, on average, because it is farther from the sun and receives only 43 percent as much solar radiation per unit of surface area. Besides this, the thin Martian atmosphere provides for very little greenhouse warming. Measurements made from the earth had shown that temperatures on the Martian equator reach 25°C at noon, but drop by 100°C or more at night. Because Mars has no ocean to buffer such changes, great swings in temperature were expected. Although no measurements of polar temperatures

---

[4] The mass of 14 micrometers of water is 0.0014 grams per square centimeter. Multiplying by the gravitational acceleration on Mars, 373 centimeters per second per second, one finds a pressure of 0.522 dynes/cm², or 0.522 microbar, where a microbar is a thousandth of a millibar.

had been made, they were not expected to be low enough to freeze carbon dioxide out of the atmosphere.

The analysis by Leighton and Murray showed that wintertime temperatures at high latitudes in both Martian hemispheres could well drop below $-128°C$, the freezing point of carbon dioxide gas under a pressure of 4 millibars. The predicted size and rate of disappearance of the polar ice caps, if assumed to be composed of solid carbon dioxide, agreed with observations of the actual Martian caps. On the basis of what was now known about Mars, there was no plausible way that the ice caps could be formed from water vapor, although small amounts of water ice were expected to be included in the carbon dioxide caps. Leighton and Murray concluded, therefore, that the Martian polar caps consist almost entirely of frozen carbon dioxide.

This prediction was confirmed in 1969, the year when two more spacecraft, *Mariner 6* and *Mariner 7,* flew by Mars. *Mariner 7* carried two infrared detectors on board as it passed over the south polar ice cap. One instrument, a radiometer, measured the heat radiated from the surface of Mars. Temperature could be deduced from these data. The other, a spectrometer, obtained both temperature data and an infrared reflection spectrum to be used for chemical identification of the constituents of the ice cap and the atmosphere. As is customary when interplanetary spacecraft are launched by the United States, the first scientific results were made public at a press conference at JPL in Pasadena shortly after the flight; the *Mariner 7* press briefing was held August 7, 1969.

Representing the team of experimenters operating the infrared radiometer, Gerry Neugebauer of Caltech reported that the maximum temperature of the icecap indicated by their instrument was $-123°C$ (later lowered to $-125°C$), a value close to that predicted by Leighton and Murray and in satisfactory agreement with the temperature expected for an ice cap of frozen carbon dioxide. The expected temperature depends on the pressure of carbon dioxide in the atmosphere—the higher the pressure, the higher the corresponding temperature, and vice versa. For 4 millibars of carbon dioxide, the expected temperature is $-128°C$; an ice cap temperature of $-125°C$ implies 6.4 millibars of carbon dioxide. These two sets of numbers were close enough to be considered in agreement.

The report of the infrared-spectrometer team, presented by George Pimentel of the University of California, Berkeley, was altogether different. Their data seemed to show, first of all, that the temperature of the edge of the ice cap was much too high for frozen carbon dioxide, and from this they deduced that at least the edge of the cap was composed of water ice. Second, it appeared that the spectrometer had detected gaseous methane and ammonia over the edge of the ice cap, although not over its main body. The presence of these hydrogen-rich gases was quite surprising on a highly oxidized planet such as Mars, and it suggested that some chemistry of an unusual kind was occurring on the planet. Putting all this together, the team proposed that a watery zone suitable for life exists at the periphery of the ice cap and that the methane and ammonia might, in fact, be biological products.

This new argument for life on Mars made headlines all over the world next day. According to the account on the front page of *The New York Times,* the findings "produced a general gasp among scientists and newsmen." Had these scientists and newsmen glimpsed the shade of Percival Lowell in the hall? Or just another false face of Mars as he misled more victims? In any event, the deception was transient. Within a few weeks, laboratory tests by the spectrometer team showed that the absorptions they had attributed to methane and ammonia could also be found in solid carbon dioxide. Their other finding, the higher temperature at the edge of the ice cap, doubtless meant that some exposed ground and rocks had come into the view of the spectrometer as the spacecraft passed onto the ice cap. Rocks and ground would, of course, be warmer than the ice cap itself. The biological interpretation was dropped, and the spectrometer data became part of the now-overwhelming evidence for carbon dioxide ice caps on Mars.

Dark Areas

The realization that the seasonally variable polar caps are composed of frozen carbon dioxide, not water ice, required a new look at the "polar seas" and other phenomena associated with the seasonal transfer of water from one Martian pole to the other that Lowell had imagined. If the ice caps are formed from the carbon dioxide always present in the atmosphere and not from water vapor moving across the Martian surface, how can one explain the seasonal phenomena?

Telescopic observations after Lowell did confirm his descriptions in a general way. The dark fringe around the disappearing ice cap is evidently real, as is the summertime increase in contrast between bright and dark regions of the planet. The wave of darkening is more open to doubt, but several observers before 1962 confirmed it, and it appears that some correlation does exist between the latitude of a region and the time at which it darkens or lightens relative to its surroundings, although this correlation is not so perfect as Lowell claimed it to be. Modern interpretations of these phenomena differ considerably from those urged by Lowell, however. The polar collar cannot be liquid water, for reasons that are explained in the next chapter. Its real cause is not known for certain, but either it may be the result of seasonal winds stripping away dust, or it may be the optical effect produced by exposure of the glassy layer of solid carbon dioxide that was mentioned earlier. Another possible source of such a layer that has not been considered is carbon dioxide hydrate, $CO_2 \cdot 6H_2O$, whose probable presence in the Martian ice cap was pointed out by Stanley Miller and William Smythe and which should form a layer at the interface between the permanent water ice cap and the seasonal CO_2 cap.

It is now generally believed that the annual changes in the dark areas result from a redistribution of dust by seasonal winds, exposing more or less of the darker bedrock. Some observers, however, have suggested that what is really involved is a brightening of the bright areas, not a darkening of the dark ones. Although the phenomenon appears to occur as a wave, whether or not this is an optical effect produced by changing illumination and viewing angles remains an open question.

Without a doubt the strongest evidence for life on Mars was that of Sinton's absorption bands. These spectral features were a modern discovery detected by reliable methods, and their apparent confinement to the seasonally variable dark areas seemed to leave little room for alternative explanations. Although NASA's advisory panel had cautioned against acceptance of an organic interpretation of the bands without further investigation, no inorganic explanation of the absorptions appeared. On the contrary, a suggestion that the bands arose from the organic compound acetaldehyde was the only proposal published before 1965 that attempted to identify these features specifically. In 1965, James Shirk, William Haseltine, and George Pimentel, all at Berkeley, demonstrated that the absorptions are better fitted by HDO (water containing an atom of deuterium, or heavy hydrogen) than by organic matter. They assumed that the HDO was located on Mars. Shortly afterward, a paper by Donald Rea, Brian O'Leary, and Sinton, from Berkeley and the Lowell Observatory, presented convincing evidence that the HDO in question was not on Mars but in our own atmosphere. (Deuterium makes up 0.02 percent of terrestrial hydrogen.) It turned out that the difficulties inherent in making spectroscopic observations on small Martian areas had led Sinton to err in assigning the absorptions he observed to the Martian dark regions. In short, there was no longer any reason to believe that the dark regions differed from the bright areas in their content of organic matter.

MARINER 9 AND PRE-VIKING MARS

Once begun, the unraveling of Lowellian Mars proceeded swiftly, and by 1969 the delowellization of Mars was complete. In place of a harsh but recognizably earthlike planet, a Mars came into view that was almost moonlike in its hostility. This new Mars had a thin atmosphere, composed predominantly of carbon dioxide, that provided little protection against solar ultraviolet radiation which penetrated to the surface of the planet almost unfiltered. Furthermore, all attempts to detect the life-essential element nitrogen, the most abundant gas of the earth's atmosphere and the one that was supposed to compose the bulk of Mars' atmosphere in the Lowellian era, had so far failed. It now appeared that nitrogen made up less than 5 percent of the Martian atmosphere, and the possibility that the planet had no nitrogen at all had to be considered. Most ominous of all from a biological viewpoint was the dryness: the low surface pressure meant that water could not exist on the surface of Mars in liquid form, but only as ice or vapor.

The television pictures returned by *Mariners 4, 6,* and *7* were just as discouraging as the atmospheric results. Mars looked more like the moon than an earthlike planet. With a few exceptions, even the familiar surface markings disappeared on close approach, and none could be identified with particular surface features. Even the boundaries between classical bright and dark regions—so clear when viewed from afar—were invisible in pictures that otherwise showed Mars in greater detail than had ever been seen before. It now appears that the bright areas consist of relatively smooth and level ground that has accumulated a more or less continuous cover of light-colored dust. The dark areas, on the other hand, correspond to steeply sloping or heavily cratered regions where the dust layer is not continuous and where the darker bedrock shows through. Of Schiaparelli's and Lowell's canals, the only traces are possible chance alignments of craters and other natural surface features that the eye connects up to form lines.

The prospects for life on Mars seemed so dim by 1970 that there seemed little good reason to emphasize biological questions in planning the spacecraft that would land there in 1976. A radical reversal took place in 1971, however, when another Mariner mission was sent to Mars. Of the two spacecraft launched in 1971, one —*Mariner 9*—achieved orbit around Mars, the goal of the mission, and it operated there for 11 months. Its most important accomplishment was to photograph the entire planet, and as large regions not previously seen by spacecraft came into view, it soon became obvious that Mars was not just another version of the moon, as earlier Mariner pictures had led one to expect, but was a planet with its own complex history.

Several spectacular discoveries led to this conclusion, among them four gigantic, inactive volcanoes, one of which is the largest in the solar system. But the aspect of Mars that has attracted the most attention is undoubtedly its multitude of channels, some of them hundreds of kilometers in length and apparently cut in the past by running water. (These channels, which are invisible from the earth, have no relation to Lowell's canals.) A number of morphologically distinct channel types are seen, but not all of them require running water to explain their origin. Some may have been eroded by glacial movement, for example, and some by flowing lava, to mention two possibilities. Many, however, and perhaps

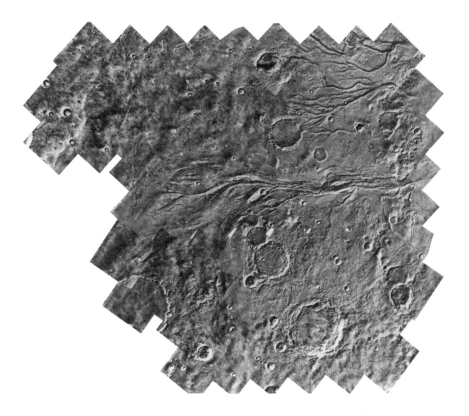

Figure 5-3   Dendritic flow channels on Mars. (If the channels appear to be hills, rotate the picture 180°.) The area shown is west of the *Viking 1* landing site. Old impact craters, formed before the flooding and eroded by it, and young craters, formed later, are visible. This photograph and the following one were taken by the Viking orbiters (National Space Science Data Center).

most, were very probably formed by water. Among these are sinuous, meandering streambeds that, with their tributaries, form typical drainage patterns. The source of water for these systems could have been subsurface ice (permafrost) that was melted by geothermal activity and seeped through the surface, but other sources—even rain—have not been excluded. Other Martian channels start abruptly as very large features, as if caused by sudden, catastrophic flooding. Unlike typical drainage systems, however, they fre-

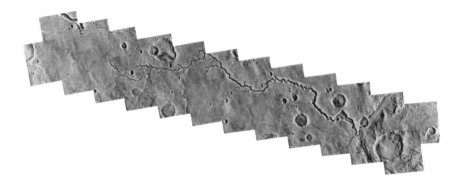

Figure 5-4 **Nirgal Vallis, in the southern hemisphere of Mars. This striking channel is 800 kilometers in length (Carr, 1981; National Space Science Data Center).**

quently diminish in size downstream. It is less certain that these channels were eroded by flowing water, although it is not impossible.

None of this streambed cutting happened in recent times. The evidence indicates that the channels are ancient—billions of years old in most cases, judging from the number of meteoritic impact craters that overlie them. Nor is there any clear evidence that lakes or oceans ever existed on Mars. The rivers did not flow into seas, but—from the evidence that remains—simply petered out, either disappearing into the ground or evaporating.

The possibility that liquid water once flowed on the surface of Mars improved the biological outlook considerably. If conditions during the early history of the planet were so temperate that water could exist on its surface, then life may have originated. If so, it was conceivable that, by adapting to gradually deteriorating conditions, Martian life had managed to survive and was still surviving on the planet. The probability that this was so did not seem very high, but in such matters, *a priori* judgments do not carry much weight when an empirical test can be made. Such a test became the major objective of the Viking mission, the next and climactic chapter in our story of the search for life on Mars.

CHAPTER SIX

The Viking Mission: Water, Life, and the Martian Desert

Water is the best thing of all.
Pindar, *First Olympian Ode* (476 B.C.)

Centuries of speculation and myth making about Mars and its inhabitants culminated in the summer of 1976 when two U.S. spacecraft named Viking arrived at Mars. The main purpose of Viking—technologically the most advanced unmanned space mission that had been flown up to that time—was to learn whether Mars actually is an abode of life. Each Viking spacecraft consisted of two parts, an orbiter and a lander, making four different units altogether. After their separation, the orbiters continued to circle the planet, photographing it and making global measurements of water-vapor distribution and surface temperature. They also functioned as relay stations for radio transmissions from the landers. The landers, after reaching the surface of Mars, performed a variety of experiments designed to answer questions about Martian biology and planetology. This chapter deals with the biologically important findings of the orbiters; the discoveries made by the landers are the subject of the next chapter.

WATER, ICE, AND WATER VAPOR

In one crucial respect, the earth is without parallel in the solar system: it is the only body orbiting the sun that has liquid water on its surface. In fact, it possesses not just traces of water, but immense quantities of it. Over 70 percent of the earth's surface is covered by oceans, and they contain enough water to submerge the entire globe to a depth of some 2700 meters, or about 1.7 miles. An observer on another planet might find it hard to believe that a place as wet as the earth contains vast areas where water is the life-limiting factor. Nevertheless, this is the case. Deserts that occupy one-fifth of the land area eloquently testify to the importance of a continuous supply of liquid water for life on our planet.

A realistic view of water on Mars did not begin to emerge from the murk that enveloped most things Martian until 1963. By 1970, five years before the Viking spacecraft were launched, earth-based and flyby observations had made it quite clear that scarcity of water must pose a major problem for any presumed Martian biota. The picture was completed by the *Mariner 9* and especially the Viking orbiters, which mapped the distribution of Martian water vapor as a function of locale and season. The findings, obtained by infrared spectrometers carried on the orbiters, showed the Martian desert to be unmitigated in its dryness. To make their meaning clear, however, we should first briefly review the physical chemistry of water.

Like many other substances, water exists in three phases—solid, liquid, and vapor—that are readily interconvertible. If a container of liquid water is left open in a room, water molecules leave the surface of the liquid and enter the air of the room, where they exist as water vapor. Occasionally, molecules of water vapor reenter the container and rejoin the liquid phase, but the traffic is predominantly in the other direction, with the result that the liquid evaporates. If the container is closed to prevent the escape of water vapor, the space above the liquid soon becomes saturated with vapor, and now the rate of condensation of vapor equals the rate of evaporation from the liquid surface; at this point, the system no longer changes and is said to be in equilibrium. The pressure of water vapor that exists at equilibrium can be measured, and it has been found to depend on the temperature. The higher the temperature, the higher the equilibrium pressure. At 25 °C, for example, the equilibrium vapor pressure is 31.7 millibars, or about 0.03 atmosphere. This means that the system is stable at 25°C as long as the pressure of water vapor in the surroundings is 31.7 millibars. At lower vapor pressures, water evaporates, and at higher vapor pressures vapor condenses, until equilibrium is attained. At 100°C, the equilibrium vapor pressure is 1013 millibars, or one atmosphere (at sea level). Bubbles of water vapor can now form in the liquid phase, and the water is said to boil.

Now let us lower the temperature to below the freezing point, so that ice replaces liquid water. Ice, too, has a vapor pressure, as shown by the fact that it evaporates in dry air. At a temperature of, say, −20°C, the vapor pressure of ice is 1.0 millibar. At −10°C, it is 2.6 millibars. In air containing lower pressures of water vapor than these, the ice evaporates, or sublimes; if the water-vapor pressure is higher, the vapor condenses directly to ice, as in the formation of frost on a cold, clear night; in either case, the transition is between vapor and solid, and liquid water does not form.

In these examples, at most two phases are present, either vapor and liquid, or vapor and ice. It is also possible, by compressing ice and thereby causing it to melt, to bring liquid and ice into equilibrium at temperatures below 0°C without a vapor phase. To put all three phases in equilibrium, it is necessary to adjust the temperature to 0°C, where the vapor pressures of water and ice are 6.1 millibars. Under these conditions, water is said to be at its "triple point." The pressure at the triple point is important for our purpose, because it is the lowest pressure at which pure liquid water is stable.[1]

Up to this point, we have described only pure water, which is rarely encountered in nature. Even rainwater contains dissolved atmospheric gases, and lake, stream, and ocean waters contain dissolved solids, as well. The presence of dissolved materials in water (or any other solvent) lowers its vapor pressure, and this in turn causes the freezing point to be depressed and the boiling point to be elevated. The magnitude of these effects depends on the

---

[1] Note that 6.1 millibars refers to water-vapor pressure, not total atmospheric pressure, as is sometimes assumed.

amount of dissolved substance. Concentrated solutions may diverge considerably, but dilute solutions differ only slightly from pure water. For dilute solutions, Raoult's law applies—the vapor pressure is proportional to the fraction of water molecules in the solution.

To cite a few examples, the vapor pressure of a sucrose solution containing 99 water molecules for every sugar molecule (a 16 percent solution by weight) is almost exactly 99 percent of that of pure water at the same temperature. The freezing point of this solution is $-1.10°C$. Seawater is a complex mixture of salts whose vapor pressure is 98 percent that of pure water at the same temperature, and its freezing point is $-1.87°C$. From Raoult's law, we conclude that 98 percent of the molecules of seawater are water molecules. (Where electrolytes are involved, as they are in this case, the ions count as molecules). The Great Salt Lake and many other saline lakes are saturated, or nearly so, with sodium chloride (NaCl, common salt). The vapor pressure of a saturated solution of NaCl is 75 percent that of pure water, and its freezing point is near $-21°C$. The fraction of water molecules in the solution is 0.82 (Raoult's law is only approximately obeyed at this high concentration). Another salt, calcium chloride ($CaCl_2$), is rare in nature, but a pond in Antarctica (described later in this chapter) is saturated with it. The freezing point of saturated calcium chloride is $-51°C$, and its vapor pressure at room temperature is only 31 percent that of pure water.

As these examples show, the addition of solutes to water stabilizes the liquid phase at vapor pressures and temperatures below those that apply to pure water. Before Viking, it seemed possible that this effect might provide a means for maintaining liquid water on Mars. We shall now consider this proposition, along with some facts about the biological usefulness of water containing high concentrations of solutes.

WATER ON MARS

The Viking Findings

The question of water on Mars—its abundance, form, and distribution—was the subject of intensive investigation both before and during the Viking mission. That Mars is a desert was known even to Percival Lowell. No one knew how dry Mars actually is, however, until the presence of water vapor in the Martian atmosphere was established by spectral evidence in 1963. The amount of water detected was 14 precipitable micrometers, equivalent to a vapor pressure at the surface of Mars of 0.5 microbar (see Chapter 5). Later observations from the earth and from pre-Viking spacecraft confirmed the identification of water vapor and showed abundances of up to 50 micrometers precipitable, or a pressure of roughly 2 microbars. (For comparison, the pressure of water vapor in the earth's atmosphere, on the equator, averages 28 millibars, or 28,000 microbars). As we have seen, a vapor pressure of at least 6100 microbars (6.1 millibars) is required to prevent evaporation of pure water, so it was clear from the beginning that liquid water —even water containing high concentrations of dissolved substances—must be very rare on Mars, if it exists at all.

The Viking investigations greatly extended our knowledge of the quantity and distribution of water in the Martian atmosphere. Data were gathered for a full Martian year, with much higher spatial resolution than earth-based observatories can achieve and over regions of Mars not accessible from the earth. In brief, the amount of water vapor detected by the Viking instruments varied from 0 to 120 precipitable micrometers (equivalent to a pressure of about 4.5 microbars at the surface), depending on the season and location. The highest abundances were found in the atmosphere over the border of the north polar ice cap, a region not observable from the earth during north Martian summer, at latitudes 70 to 80°N, in midsummer when the ice cap had shrunk to its "residual," or minimum size. The residual ice cap consists of water ice, as shown by the abundance of water vapor in the atmosphere above it and also by the temperature of the cap itself: $-68°C$ in the summer of 1976, the year of the Viking landings—much too high a temperature for an ice cap of solid carbon dioxide.

As the Viking sensors moved southward from the latitudes of peak water-vapor abundance, the quantity of water in the atmosphere dropped sharply, reaching its lowest values in the southern hemisphere, as shown in Figure 6-1. The evidence leaves little doubt that the north polar region is the major source of atmospheric water on Mars during northern summer.

With the approach of fall and winter in the north, atmospheric

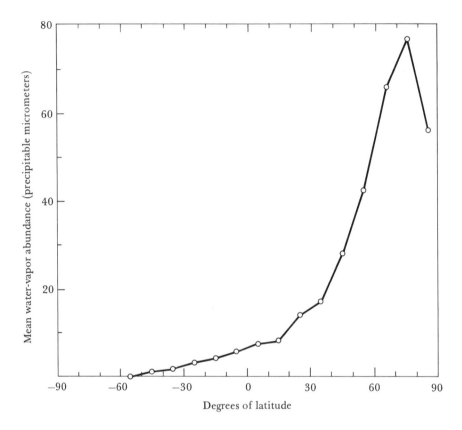

Figure 6-1   Martian atmospheric water vapor as a function of latitude during northern summer. Observations were made at longitude 180°W between 1200 and 1400 hours Martian local time. (Redrawn from Farmer et al., 1976; copyright 1976 by the AAAS.)

water-vapor abundances declined in the northern hemisphere but increased only moderately in the south. The peak values seen during northern summer were not repeated in the southern hemisphere in its 1977 summer season. The temperatures of the south polar ice cap remained close to the freezing point of carbon dioxide all summer, and therefore the ice cap could not have served as a source of atmospheric water vapor, even if it contained water, which is likely but not certain. The marked difference between the two ice caps displayed in the Viking measurements is thought to be caused, in part, by dust storms that originate uniquely in the southern hemisphere during southern summer. One of the effects of the suspended dust is to screen the southern ice cap from the heat of the sun.

At equatorial latitudes, where ground temperatures often exceed 0°C (and which for this reason seemed, in the pre-Viking era, to be particularly favorable latitudes for life), water-vapor abundances remained at 5 to 15 precipitable micrometers throughout the year. With so little water vapor in the air, the surface must be exceedingly dry. In fact, C. B. Farmer and P. E. Doms have concluded from the Viking data that the entire region between 35°S and 46°N is a desiccated zone that has lost whatever water it may have had to the polar regions, which act as cold traps. It is considered likely that large amounts of water are retained in the Martian circumpolar regions as subsurface deposits of permanently frozen ice, or permafrost. This permafrost surfaces at the north pole, where it forms the residual ice cap—the tip of the iceberg, so to speak. It is possible, but not yet demonstrated, that a permanent deposit of water ice comes to the surface at the south pole, as well.

This state of affairs differs sharply from that found on the earth, where atmospheric water vapor is most abundant at the equator and least abundant at the poles. On our planet, average water-vapor pressure at 0° latitude is 28 millibars; at 70° it is 1.3 millibars, averaged over both hemispheres. These differences exist because water is found at all latitudes on the earth, and the amount of water vapor in the atmosphere is largely determined by the temperature, which is high at the equator and low at the poles. On Mars, water (in the form of ice) is confined almost entirely to the polar regions where the low temperature of the atmosphere severely limits the amount of water vapor it can hold. Even when the Martian atmosphere is saturated with water vapor, the vapor pressure is extremely low.

The Viking findings confirmed what had been foreshadowed by the earlier observations: water-vapor pressures everywhere on Mars fall far short of the level required to maintain liquid water on the planet. In fact, the Viking data showed that Mars is even drier than it had earlier appeared. Before Viking, it was thought that the water vapor of the atmosphere was close to the ground and settled

Figure 6-2 This 85-degree panorama of the *Viking 2* landing site was taken early in the mission, during northern summer (National Space Science Data Center).

out at night as frost. It seemed possible that the frost could melt after sunrise, producing briefly moistened soils that might conceivably sustain a microbial population. A theoretical analysis of this problem by Andrew Ingersoll in 1970 showed, however, that owing to the low temperature of Mars and the composition and low pressure of its atmosphere, frost on the surface would evaporate before it could melt. C. B. Farmer then argued that melting could occur if, after its deposition, the frost were covered over by enough fine, wind-borne dust to retard the escape of water vapor.

This discussion now seems academic. The Viking results showed (1) that there is very little water vapor anywhere in the Martian atmosphere, and (2) that the vapor that is present is not concentrated near the ground but, in most seasons and places, is distributed throughout the atmosphere to a height of 10 kilometers or more. Under these conditions, no significant nightly deposition of frost could occur. Although nightly ground fogs composed of tiny ice particles were observed by the cameras on both Viking landers, the particles were too small to fall to the ground.

While these discoveries appear to rule out a daily condensation of significant quantities of atmospheric water vapor onto the Martian surface, a seasonal transfer of water from the atmosphere to the ground and back again clearly does take place, at least in the north polar region. Cameras on *Viking Lander 2,* the more northerly of the two landers, observed a thin layer of frost that appeared on the ground around the craft and remained for several months during two winter seasons. This frost could not have been a direct atmospheric condensate because there was too little vapor in the at-

Figure 6-3   This photograph, taken after the start of winter in the north, shows the left-hand portion of the scene in Figure 6-2. Much of the surface is now coated with a thin layer of frost (National Space Science Data Center).

mosphere at the time to condense. It has been suggested that the frost originated in the southern hemisphere and was carried on dust grains to the north polar region where CO_2 condensed on them and made them heavy enough to fall to the ground. The CO_2 then evaporated, leaving the water ice behind. A significant net movement of water from the southern hemisphere to the north appears to take place by this or some other mechanism, but most of the annual condensate in the arctic region is water that moves back and forth seasonally between the ground and the atmosphere.

Martian Salt Pools?

Let us now consider the idea that liquid water might exist on Mars in the form of highly concentrated salt solutions. An especially effective salt for this role, if it exists on Mars, would be calcium chloride. At its freezing point of $-51\,°C$, a saturated solution of calcium chloride has a vapor pressure of 34 microbars. The maximum pressure of water vapor in the Martian atmosphere is only about 4.5 microbars, however, so that a saturated solution of calcium chloride would slowly but surely evaporate. To maintain the solution, the water would have to be replenished from time to time, and the seasonal deposition of frost in the polar regions might conceivably meet this requirement. Temperature measurements at the *Lander 2* site show, however, that such a solution would be frozen all winter and could thaw only in the daytime in the summer.

Calcium chloride is probably rare on Mars for the same reasons that make it rare on our planet. On the earth, calcium is found chiefly as limestone (calcium carbonate) and gypsum (calcium sulfate). Both of these salts are much less soluble than the chloride, and they precipitate out of solution ahead of it. Mars has an abundance of carbon dioxide in its atmosphere and sulfate in its soil, as the Viking inorganic soil analysis indicated, and it is likely that calcium carbonate and sulfate formed wherever surface water existed in the past. No other salts that are reasonable for Mars would be useful for maintaining liquid water on the planet.

Life at Martian Temperatures

The low temperature of Mars is clearly a major factor governing the state of water on the planet. The average temperature of the Martian surface is $-55\,^{\circ}C$, compared to $15\,^{\circ}C$ for the earth (Table 4-1). Even on the equator, Martian nighttime temperatures drop far below zero, although noontime readings up to $25\,^{\circ}C$ have been obtained. While harsh by terrestrial standards, however, Martian temperatures do not by themselves exclude the possibility of life on the planet. Growth of terrestrial microorganisms is well documented for temperatures down to $-10\,^{\circ}C$, and one report cites growth of a yeast at $-34\,^{\circ}C$. Survival without growth has been reported for a variety of cells at temperatures as low as $-196\,^{\circ}C$. It seems reasonable to assume, then, that if a suitable solvent existed on Mars, Martian temperatures would not preclude active life, not, at least, in some regions of the planet.

Conclusions

It is very unlikely that liquid water of any kind exists on the Martian surface, even transiently. Martian life, if any, must by its own efforts produce liquid water from atmospheric water vapor or from ice for use as a solvent. These processes require the expenditure of considerable energy. Some desert-dwelling terrestrial organisms do use water vapor to obtain water in dry environments. The following section describes some of the means by which desert dwellers deal with the water problem.

BIOLOGICAL WATER REQUIREMENTS

Water Activity

All cells (except those in a dormant state) live in an aqueous solution of some kind. The cells of higher animals live in blood plasma, those of plants in tissue sap, and free-living cells such as bacteria live in a variety of aqueous environments. Plants and animals create an internal environment for themselves; the cells of microorganisms, however, cope directly with the outside environment.

The water requirements of cells are conveniently discussed in terms of the water activity of the medium in which they live. Water activity, abbreviated a_w, is a measure of the effective concentration of water in a solution—that is, the concentration of water available for chemical reactions. In any aqueous solution, some of the water is combined with the molecules or ions of the solute in complexes called hydrates. It is the formation of hydrates that brings the solute into solution. Because those water molecules that are involved in hydrate formation are not available for other reactions, the water activity of a solution is lower than that of pure water. The vapor pressure of a solution, which is directly related to water activity, is also lower than that of pure water. In fact, water activity is defined as the ratio of the vapor pressure of the solution, p, to that of pure liquid water, p_0, at the same temperature:

$$a_w = \frac{p}{p_0}$$

The water activity is numerically identical to the relative humidity of air in equilibrium with the solution. Thus, a saturated solution of sodium chloride ($a_w = 0.75$ at $25°C$) enclosed in a small space will bring the air in the space to a relative humidity of 75 percent. In dilute solutions, water activity is equal to the fraction of water molecules in the solution, a statement that follows from Raoult's law.

Higher Plants and Animals

The water activities of some solutions of biological interest are shown in Table 6-1, recalculated where necessary from other measures of active water concentration. All multicellular species require high water activities for growth and metabolism. Human blood plasma, the medium we live in, is typical for mammals. From the standpoint of its water activity, it is barely distinguishable from distilled water. Its activity is, in fact, matched by 0.9 percent NaCl, a faintly salty solution known as "physiological saline." The cell sap of most plants is similar in its water activity to the blood of animals.

Desert plants. One might suppose that the cells of plants and animals adapted to life in dry environments would have lower requirements for water than those of other species, but this is not the case. Desert-dwelling species survive by employing elaborate stratagems that enable them to maintain their internal fluids at water activities that differ little from those enjoyed by species from wet environments. In plants, the most important tactic consists simply of storing water. Most deserts receive occasional rain, and plants such as cacti and other succulents accumulate water in their stems or leaves in wet weather and hold it for use during dry periods. In addition, these plants reduce the rate of water loss from their leaves and stems by closing the stomata (pores) through which gas exchange normally occurs. Since photosynthesis depends on gas exchange with the atmosphere, this closure has the additional effect of slowing the growth rate. According to P. S. Nobel, the water activity of cellular fluids in only a few desert plants drops as low as a_w 0.96 at 25°C.

Other desert plants do not conserve water but go through their entire life cycle during the short season when water is available, leaving only dormant seeds or bulbs to survive the subsequent drought. The dormant cells presumably do come into water equilibrium, or nearly so, with the external environment. Dormancy is also a response of some perennials to extreme drought.

For such plants, survival depends on abundant, if infrequent, supplies of liquid water in the form of rainfall. Another potential source of water is atmospheric water vapor, which is present in even the driest terrestrial deserts in amounts that are enormous by

Table 6-1   Water Activities of Some Solutions of Biological Interest

Solution	Temperature (°C)	a_w	Source
Water	All	1.000	
Human blood plasma	37	0.994	Schmidt-Nielsen
Cell sap, peas	25	0.994	Nobel
Dipodomys blood plasma	37	0.993	Schmidt-Nielsen
Tenebrio body fluids	25	0.987	Edney
Seawater	25	0.98	Sverdrup et al.
Saturated sucrose	25	0.85	Robinson and Stokes
Saturated NaCl	25	0.75	Robinson and Stokes
Saturated $CaCl_2$	25	0.31	Weast
Saturated $CaCl_2$	0	0.42	Weast

Martian standards. Relative humidities are typically low in the daytime, when air temperatures are high, but many desert atmospheres become saturated when the temperature falls at night, allowing liquid water to form as dew or fog. These are important sources of water for some plants, but with the possible exception of the Chilean shrub *Nolana mollis,* there is no established example of a higher plant that utilizes water *vapor* for photosynthesis and growth. Although water vapor taken up by the leaves of some plants—bromeliads, for example—probably contributes to their survival, it is difficult for a plant to absorb enough water vapor from the atmosphere to make growth possible, since even saturated air contains very small quantities of water. While one gram of water in liquid form occupies a volume of 1 cubic centimeter, for instance, the same amount of water as a vapor occupies 43,500 cubic centimeters in air saturated at 25°C. Needless to say, the acquisition of water vapor by plants is a very slow process.

One shrub that grows in the Atacama Desert of northern Chile, *Nolana mollis,* has been found to condense water vapor out of the atmosphere by secreting salts (largely NaCl) through special salt glands in its leaves. Water vapor condenses on the leaves when its pressure in the atmosphere exceeds that of the saline secretion. Although the relative humidity in this region seldom rises above 80 percent, even at night, sufficient condensate forms to drip off the leaves and wet the ground. It seems likely, although it has not been proven, that on occasions of high humidity, when enough water has condensed to dilute the salts below a certain critical value, the saline solution can be taken up by the roots, the salts removed, and the water used by the plant. A similar mechanism has evolved in insects, as we shall see.

Certain desert lichens have been shown to use water vapor for photosynthesis. Like other microorganisms, lichens do not create a constant internal environment for their cells, which must survive under the changing conditions of the world around them. In the case of desert lichens, this means existing in a dry, dormant state for much of the time but becoming activated rapidly on contact with moisture. Desert lichens ordinarily depend on fog and dew for water, but some species can achieve net photosynthesis (an excess of photosynthesis over metabolic decomposition of carbohydrate) using water vapor. Lange and his colleagues have shown that *Ramalina maciformis,* a lichen of the Negev Desert, can photosynthesize when the relative humidity exceeds 80 percent. Antarctic lichens have also been found to use water vapor for photosynthesis. Growth is predictably very slow when water vapor is the only moisture source.

Desert animals. Animals that inhabit arid environments do not store water and sometimes do not even drink it. Rather, they manufacture and conserve it. (Contrary to common belief, the camel does not store water. It can withstand considerable dehydration, however, and it can build up and survive large water deficits.) One of the most interesting nondrinkers is the kangaroo rat, *Dipodomys merriami,* a small rodent of the Arizona and California deserts, which has been described in a fascinating account by Knut Schmidt-Nielsen. The kangaroo rat does not normally drink water, even when it is available. It produces virtually all of its water by oxidizing the organic matter (chiefly carbohydrates) contained in the seeds and dried plant materials that make up its diet. All aerobic organisms necessarily produce water in the course of their metabolism, but the ability to survive on metabolic water alone is limited to a small number of animals that have evolved special physiological and behavioral mechanisms for this purpose.

In the case of the kangaroo rat, the animal is nocturnal; during the heat of the day, it remains in its underground burrow where the temperature is relatively low and the humidity high. It further reduces the loss of its body water by having no sweat glands, by producing very concentrated urine and dry feces, and by expiring little water in its breath. In laboratory experiments, Schmidt-Nielsen found that kangaroo rats can live indefinitely without water on dried barley at a relative humidity as low as 24 percent. At 10 percent relative humidity, the animals lost weight, showing that at this point and below they were unable to maintain water balance. When they were fed soybeans instead of barley and conditions were normally humid, they produced so much urea (because of the high protein content of the beans) that they had to drink water to maintain their water balance. So powerful are their kidneys, however, that the animals were able to drink seawater!

As Table 6-1 shows, the kangaroo rat's adaptations to desert life do not include any reduction in the basic water requirements of its cells: the water activity of *Dipodomys* blood is virtually the same as our own. It is important to understand that animals that live without drinking water are not going without water. They are actually consuming the water used for photosynthesis by their food plants. The carbohydrate that *Dipodomys* converts into water is, in effect, a stored form of water (see Equation 4-2).

Many insects that inhabit dry environments—the mealworm beetle, *Tenebrio molitor,* that infests flour and grain, for example—live on metabolic water. Beside using metabolic water, *Tenebrio* and some other insects also use water vapor that they are able to extract from unsaturated atmospheres. The mechanism that *Tenebrio* employs involves the creation of a concentrated salt solution in tubules associated with the hindgut, into which water vapor is absorbed through the gut wall. Water can be obtained in this way from atmospheres with relative humidities down to 88 percent. Since the vapor pressure of the insect's body fluids (shown in Table 6-1) is considerably higher than this, *Tenebrio* must expend energy to acquire such water. Other insects have been said to remove water vapor from the atmosphere at relative humidities as low as 45 percent. In such cases, solutions of very soluble organic compounds, not inorganic salts, are presumably used to condense water vapor.

Microorganisms

Life in brines and syrups.   Like other cells, microorganisms live only in aqueous solutions except when dormant, but many can function at water activities far below those required by the cells of higher plants and animals. Even so, the majority of microbial species require water activities of 0.90 or higher in order to grow. This fact underlies the prescientific discovery that meat and fish can be preserved by drying, or by salting, which amounts to the same thing. Fruit is preserved in saturated solutions of sucrose, as in jam. Occasional spoilage of such products shows, however, that some organisms can grow at water activities of 0.85 and even 0.75 (see Table 6-1). The lowest water activity at which microbial growth has yet been reported is at a_w 0.61. A mold, *Xeromyces bisporus,* and a yeast, *Saccharomyces rouxii,* grow slowly in sugar solutions at this water activity, although both organisms prefer more water. *Xeromyces,* for example, grows optimally at a_w 0.92. The capacity for growth at a_w 0.75 or lower is found not only in yeasts and molds, but also in some bacteria and algae.

Do microorganisms perhaps escape the low water activity of their environment by maintaining a high water activity inside their cells where they carry out their chemistry? The answer is that they do not. Cell membranes are too permeable to water to make this a practical solution. Instead, these organisms have learned to live at, or close to, the water activity of the medium. (The review by A. D. Brown in the Bibliography describes the various physiological adjustments that make this possible.)

Life in the Antarctic desert.   Biological exploration of one of the world's most extreme deserts (and the only one that begins to approach the rigors of the Martian environment to any degree) became possible following the International Geophysical Year of 1957–1958, which drew attention to the remote Antarctic continent. One result of the IGY was an international treaty, ratified in 1959 by sixteen nations, recognizing Antarctica as a nonmilitarized zone reserved for scientific studies for a period of thirty years.

Antarctica is known to be covered by a great ice cap, and many people were surprised to learn, after the IGY, that ice-free regions exist on the continent. The largest of these, a cold desert several thousand square kilometers in area and usually referred to as the "dry valleys," is located near McMurdo Sound in southern Victoria Land. Low temperatures and scarcity of liquid water are the predominant ecological features of these valleys. The mean annual air temperature is about $-20°C$, and the summer average is close to $0°C$. Precipitation is sparse—approximately 10 centimeters (4 inches) of water per year, all in the form of snow. The low precipitation reflects the limited water-holding capacity of the cold atmosphere. Cut off from the general flow of glaciers from the interior of the continent by the Transantarctic Mountains, the region is ice-free, and it is swept by strong and cold, but very dry, katabatic winds that blow off the high Antarctic plateau. The winds cause the snow to evaporate, with little melting.

There is reason to believe that the valleys have been drying out for thousands of years. Saline lakes and ponds, fed during the short summer by trickles of glacial meltwater, have no drainage, since the ground is frozen, yet they are smaller than their basins. The difference is due to loss by evaporation. Some of the lakes have terraces with dead algal mats marking past water levels. Carbon dating of the mats has shown that 3000 years ago the surface of Lake Vanda, for example, was 56 meters above its present level. The age of the whole dry-valley system has been estimated at between 10,000 and 100,000 years.

Life in the valleys is almost entirely microbial. Dense populations of algae and cyanobacteria (formerly called blue-green algae) are found on the shorelines, and these photosynthesizing species support large populations of bacteria, yeasts, and molds. Microscopic animals—protozoa, rotifers, and tardigrades—are also found. The organism count declines markedly when one leaves the streams and ponds. Even lichens, the most successful of Antarctic land organisms, are absent at the dry, higher elevations of the valleys. William Boyd, one of the earliest workers in the field, reported that some soils from the driest locations contained no detectable bacteria. Later teams of biologists have obtained the same result. Roy Cameron, a soil microbiologist from the Jet Propulsion Laboratory, examined hundreds of samples in the course of eight seasons in the dry valleys. About 10 percent of these samples contained no detectable organisms, and many other samples showed very low microbial counts. Robert Benoit and Caleb Hall had similar results. "At those soil sites which receive the least amount of water," they wrote, "the surface inch of soil was often abiotic (i.e., no detectable life) or had less than 10 bacteria per gram of soil." Deeper samples usually had measurable populations, but at one site Benoit and Hall detected no bacteria from the surface to a depth of one meter.

The fact that these soils support growth if moistened shows that water is the life-limiting factor in dry-valley soils. Low temperature is not a dominating factor, and, in fact, many of the microorganisms of the region, especially those at the lower and wetter elevations, can grow and photosynthesize at temperatures near 0°C. On the other hand, tolerance for low water activities is rare among the organisms found in the valleys, despite the abundance of saline habitats, soils as well as ponds. This and other evidence suggests that the small microbial populations of the dry soils are not indigenous to the region but are carried in by winds from more favorable locations. Such organisms find favorable conditions only in protected microenvironments in the valleys. Thus, Cameron observed algae growing on the undersides of translucent pebbles where they are protected from desiccation, and Imre Friedmann has shown that bacteria and lichens can live within translucent, porous rocks, under north-facing surfaces that receive enough sun to melt snow, which is then absorbed by the rock.

Failure of the microbial life of the valleys to cope with the effects of this pervasive dryness is also seen in the Don Juan Pond, a shallow body of water of some 10 to 20 acres that comes as close as anything on our planet to the hypothetical (and probably nonexistent) Martian puddle (see Plate 8). At the time of its discovery in 1961 the pond was not frozen, although its temperature was −24°C. Different observers have subsequently measured its freezing point at from −48 to −57°C. The pond is saturated with calcium chloride, which is crystallizing as the hexahydrate $CaCl_2 \cdot 6H_2O$. These crystals, which were known only in the laboratory before they were identified in the Don Juan Pond, have been given the mineralogical name antarcticite. Their formation is made possible through the combined effects of the very low temperatures and the extreme aridity that characterize the dry valleys.

Microorganisms living around the two freshwater inlets that feed the pond are washed into it from time to time, so it would not be surprising to find occasional organisms in the pond. An early report of microorganisms growing in the pond—at a water activity of about 0.40 (Table 6-1)—was unexpected, however, and has not been confirmed by later investigators. The pond appears to be essentially sterile, a result consistent with the observation that tolerance to high salt concentrations is rare among microorganisms of the valleys.

Contrary to the usual view, these Antarctic studies seem to say that the adaptability of life is not almost limitless. Rather, it appears that the conditions under which life can survive are actually quite narrow.

Conclusions

Although the adaptations of desert organisms to life with limited water are ingenious and often surprising, they become almost irrelevant when measured against the extreme dryness of Mars. Among all known terrestrial species, only water-vapor utilizing lichens can even be considered as possible models for Martian life. All the others require liquid water directly or indirectly. The insects just described fall into the latter category: water vapor merely supplements their main water supply, which is the carbohydrates of plants. The lichens have not been observed to utilize water vapor at relative humidities below 80 percent in warm deserts, and they have not colonized the Antarctic dry valleys, which are very wet by Martian standards. It appears that if life exists on Mars, it must operate on different principles from terrestrial life with regard to use of water.

ANTARCTICA AND THE MARS QUARANTINE POLICY

The first reports of sterile soils in Antarctica were met with universal scepticism. To speak of "sterile soil" is to commit a microbiological impropriety—every biologist knows that microorganisms are an essential component of what is usually understood by the word "soil," which is simply the material that plants grow in. Since plants do not grow in the dry valleys, it is perhaps arguable whether the surface material there should even be called soil. In any case, it was only after investigators had accumulated much evidence that the unsuitability for life of the dry-valley desert could be taken seriously.

My colleagues and I advanced this idea, but agreement was by no means unanimous. The view was unconventional, but beyond that, its implications for the Mars explorations were bound to be disputed. Specifically, it cast doubt on the idea that Mars was vulnerable to contamination by terrestrial bacteria. This widely held belief, which had its roots in Lowellian Mars, underlay a large spacecraft sterilization program undertaken by NASA in response to a treaty that bound the United States and other nations to conduct space exploration in a manner that would avoid "harmful contamination" of extraterrestrial bodies. NASA had chosen to fulfill the treaty by subjecting fully assembled spacecraft intended for Mars landing to sterilization by heat. Since this procedure would add considerably to the cost of exploring Mars and might, in addition, damage both spacecraft and instruments, the basic assumptions of the quarantine policy and the details of the sterilization procedure became matters of considerable debate after 1963, when the true nature of the Martian environment began to emerge. In this context, the message of the Antarctic findings was clear enough: if terrestrial microorganisms were unable to colonize the dry valleys—which must be Paradise, compared to Mars, for any terrestrial bacterium or yeast—there was certainly no need to worry about their overrunning Mars.

One scientist who did not accept this conclusion was Wolf Vishniac, professor of microbiology at the University of Rochester and a member of the Viking biology team. To Wolf, the idea of sterile soils, even in Antarctica, was implausible. As one who had played a leading role in setting the Mars quarantine policy, he considered the issue so urgent that he decided to go to the dry valleys himself during the southern summer of 1971–1972. He was convinced that Antarctic soils contained enough water for microorganisms, that the problem was a methodological one, and that, by applying the right procedures, actively growing microbial populations could be found in all dry-valley soils. To this end, he employed some novel methods for detecting soil organisms, with results that convinced him that he was on the right track. The work could not be completed in one season, however. He returned to the Antarctic for more fieldwork in 1973, and there he met his death in a tragic fall.

The work that Wolf Vishniac started is still unfinished. Given more time, he would have sampled more sites, and certainly he would have compared his methods with those of his predecessors on the same soil samples, a critical test. Because he could not do these things, the question of life in the dry valleys may still be an open one for some. The relevance of the answer for Mars, however, is no longer at issue. The Mars question was settled by the Viking mission. The two fully sterilized Viking landers that went to Mars found a planet that was even more hostile than it appeared to be from orbit. Mars turned out to be contamination-proof, not only because its aridity and cold far exceed anything known in Antarctica—this was clear before the Viking mission—but also because its chemistry makes it self-sterilizing, as the next chapter shows.

CHAPTER SEVEN

The Viking Mission:
Where are the Martians?

*Those who have never seen a living Martian can
scarcely imagine the strange horror of its
appearance.*

H. G. Wells, *The War of the Worlds* (1898)

In the early 1970s, the Soviets had placed three spacecraft on
Mars, but none of them returned usable data. The Viking landers
were therefore not the first to reach the Martian surface, but they
were the first to survive to tell what they found. This chapter describes the biological findings of those historic landings.

THE LANDINGS

On arriving at Mars in mid-June and early August of 1976, the
Viking spacecraft were placed in predetermined orbits around the
planet, and the search for landing sites began. Two criteria guided
the choice of sites: lander safety and scientific interest. Scientific interest had led to an earlier decision to set the spacecraft down at
widely separated latitudes that would present different Martian

climatic and geological domains. Cameras and infrared sensors on board the orbiters, along with radar from the earth, performed the search for specific locations. Sites of higher-than-average temperature and moisture were sought because they would presumably be favorable habitats for Martian life. No such places were found. In the end, there was little choice, biologically, between one locale and another at a given latitude, and the final decisions on landing sites were based on safety considerations alone.

When all was ready, a command from the earth activated the explosive bolts that held the lander to its orbiter, compressed springs separated the two spacecraft, and the lander (enclosed in a protective aeroshell) fell toward Mars from an altitude of 1500 kilometers. At 6 kilometers above the surface, as determined by the lander's radar, a parachute was deployed to slow the fall, and the aeroshell was discarded; at 1.5 kilometers, the parachute was jettisoned, the three legs of the lander were extended, and retroengines were fired for the final deceleration. Lander 1 touched down on the Martian surface on July 20 at 22.5°N, 48°W, in the region of Mars called Chryse Planitia. Lander 2 arrived on September 3 at 47.5°N, 226°W, in Utopia Planitia, some 1500 kilometers north of the latitude of *Viking 1* and on the opposite side of the planet. The season in the northern hemisphere of Mars was early summer. The mission was planned to last for 90 days after touchdown, but all four spacecraft operated for two years, and Lander 1 survived on Mars for over six years.

The use of retrorockets for the final descent to the Martian surface was worrisome because of the physical and chemical changes at the landing sites that the jets were expected to produce. In order to minimize such effects, the conventional retroengines were redesigned to lessen surface heating and abrasion, and specially purified hydrazine (N_2H_4), which does not form organic products on ignition, was used as the retrofuel. The exhaust consisted of nitrogen, hydrogen, and ammonia in roughly equal volumes, along with 0.5 percent of water vapor. Laboratory simulations of a Viking landing using this fuel and the redesigned engine showed little or no killing of soil bacteria at the simulated landing site. Significant contamination of the soil by ammonia was observed, however, and contamination by the water vapor of the exhaust also presumably occurred although it was not measured. The possibility that the ammonia and the water caused chemical changes in the soil[1] at the real Martian landing sites is considered in due course.

COMPOSITION OF THE ATMOSPHERE

As the two landers fell to the surface of Mars, instruments mounted on their aeroshells carried out pressure, temperature, and atmospheric composition measurements. These measurements were later repeated on the surface with additional instruments carried by the landers. The most important question from a biological point of view concerned the composition of the atmosphere, and especially whether or not nitrogen was present. Nitrogen is a major element of nucleic acids and proteins, and it is generally assumed to be essential for life. Earlier space missions had established that nitrogen makes up less than 5 percent of the Martian atmosphere, compared to the earth's 77 percent, and the Martian value might actually be 0 percent.

The Viking instruments found 2.7 percent nitrogen in the Martian atmosphere, together with carbon dioxide (95 percent), argon (1.6 percent), oxygen (0.13 percent), and smaller amounts of carbon monoxide, neon, krypton, xenon, ozone, and water vapor. It is likely that much more nitrogen was once present, but it has been lost to space. Although nitrogen is too heavy to escape from Mars unassisted, Michael McElroy has shown that chemical reactions at the top of the Martian atmosphere can boost nitrogen atoms so that they escape. Consistent with an escape theory, Viking found that the ratio of the heavy isotope of nitrogen, ^{15}N, to the common isotope, ^{14}N, is higher in the Martian atmosphere than in that of the earth, suggesting preferential loss of the lighter isotope from Mars.

---

[1] To avoid circumlocution as well as any chance of confusion, I shall use the word "soil" to refer to the surface material of Mars, even though it is not, properly speaking, a true soil.

THE SEARCH FOR LIFE

The Instrument Teams

The search for evidence of Martian life was carried out by six instruments that were the same on each lander: two cameras, a gas chromatograph–mass spectrometer for identifying organic compounds in the Martian soil, and three instruments designed to detect the metabolic activity of soil microorganisms. Responsibility for the operation of each instrument and the interpretation of its data was assigned to a group of scientists who were familiar with the design and capabilities of that instrument, an arrangement that held for all the instruments on the Viking landers and orbiters. These various teams of specialists, together with the engineers who operated the spacecraft and the managers who coordinated all the activities, made up the entire Viking team.

The Cameras

Among the life-seeking instruments on board the Viking craft, the cameras were unique in two respects. First, they made no assumptions as to the chemical nature of Martian life. Observers on earth had to decide on the basis of appearances alone, not physiology or chemistry, whether or not a scene showed evidences of life. A silicon Martian had the same chance as a carbon one of being picked up by the cameras. Aside from actual living creatures, the cameras could also detect such signs of life as tracks, remains, artifacts, and movement. Nothing smaller than several millimeters could be resolved by the cameras, however, and this set a lower limit on the size of interpretable objects. As we know, whole worlds of life can exist below this limit, but these were beyond the range of the Viking cameras.

The second feature that distinguished the cameras from other life-seeking instruments was that one shot, a single result, could do the whole job. Each photograph contained so much information, in the technical sense of the word, that, in principle, it was possible to prove the existence of life on Mars by a single picture. None of the other instruments had the capability to acquire completely convincing evidence for Martian life in a single observation.

Figure 7-1 A 100-degree panorama of the *Viking 1* landing site. The picture is divided by the meteorology boom of the spacecraft. The large rock on the left, referred to throughout the mission as "Big Joe," is about 3 meters in length (NASA/ JPL).

The cameras electronically registered the scene on a tape recorder in the lander. The pictures were then transmitted to the earth either directly or by relay via the orbiters. Direct transmissions without intervention of the recorder were also possible. These were still pictures, in color or black-and-white, of high or low resolution, and some sequences were in stereo.

The photographs that came back from Mars were examined carefully by many members of the Viking team for a variety of purposes, and they were scrutinized intensely for signs of life by a special subgroup of the lander-imaging team. They were studied monoscopically, stereoscopically, and in color; they were subjected to special computer techniques designed to reveal movement or changes in the scene; and they were searched for objects that emitted light at night. Nothing suggesting Martian life was encountered that did not have a more plausible nonbiological explanation. To quote the report of the special study team, "No evidence, direct or indirect, has been obtained for macroscopic biology on Mars" (Levinthal et al., 1977).

Even though the cameras saw no life, their photographs are priceless, not only for what they added to our understanding of the Martian environment, but also because these ochre landscapes from the plains of Mars will always be reminders of the historic encounter between myth and technology that took place there in the summer of 1976.

Figure 7-2   Two photographs of Big Joe (see Figure 7-1) and the surrounding area taken 25 months apart under similar lighting conditions. The later picture shows a change in the scene (labeled B) produced by the slumping of an unstable dust layer. The movement was probably caused by diurnal heating and cooling of the surface. A similar change (A) is visible on Whale Rock in the background (National Space Science Data Center).

The Gas Chromatograph–Mass Spectrometer (GCMS)

The GCMS was not, strictly speaking, a life-seeking instrument. Its role was, rather, to detect and identify organic compounds in the surface of Mars. Although virtually all organic matter found on the earth is biological in origin, nonbiologically produced organic material is abundant elsewhere in the universe, as we saw in Chapter 3. In pre-Viking times, it had been considered likely that, in the absence of life, at least meteoritic organic matter would be found on Mars. The reason for this assumption was that Mars is close to the asteroid belt, lying between the orbits of Mars and Jupiter, from which meteorites originate. Meteorite collisions with Mars are thought to be significantly more frequent than they are with the earth, and estimates suggested that sufficient meteoritic organic substance would have accumulated on Mars over geologic time to be detected by the GCMS. If, in addition, Mars were a habitat of life, then organic matter of biological origin would be present as well. One subject frequently discussed among scientists in the years before the Viking launches was whether it would be possible, with the available instruments, to discriminate between biological and nonbiological sources of the organic compounds that most of us expected to find at least traces of on Mars. As it turned out, after the landings the question never arose.

The GCMS operation involved several basic steps. The Viking surface sampler—a scoop on the end of an extensible boom—took up a small sample of Martian soil which was then ground, sieved, and delivered to a small oven with a capacity of about 200 milligrams. The function of the oven, which could be heated in steps to 500°C, was to volatilize the organic material in the sample. Small, neutral organic molecules are volatilized at low temperatures; large or polar molecules are broken down (pyrolyzed) at high temperatures to yield small, volatile fragments. The different gases formed in the oven then entered a gas-chromatographic column, a long tube packed with grains of a synthetic organic material, through which they moved at different speeds. As the mixture, now separated into its component gases, emerged from the end of the column, these were transferred one at a time into the mass spectrometer. There, each one was further decomposed, this time into charged fragments, by an electron beam, and the mass of each fragment was measured by accelerating it through electrostatic and electromagnetic fields. The resulting pattern of molecular masses contains information that enables an experienced mass spectroscopist to identify the molecule from which the fragments originated. The next step is to determine what the original molecule was that gave rise to the pyrolysis products identified in the mass spectrometer. This, too, is possible, although somewhat uncertain.

Gas chromatography–mass spectrometry was the method of choice for the Viking mission. It has important advantages over the usual methods for carrying out organic analysis: among them, it works for essentially all organic compounds. This meant that the classes of compounds that would be sought on Mars did not have to be chosen in advance. (The GCMS was, in fact, not limited to organic compounds. By bypassing the pyrolysis–gas chromatography section of the instrument, it was also used for atmospheric analysis.)

The GCMS was extremely sensitive for organic compounds. It could detect a few parts per billion of substances containing more than two carbon atoms and a few parts per million of those with one or two carbon atoms, concentrations of organic molecules 100 to 1000 times less than those encountered in desert soils on the earth. Two soil samples were analyzed at each landing site, including a sample taken from under a rock at the Utopia site, and the results were uniformly negative. The only organic materials found were traces of cleaning solvents left over from the manufacture of the instrument. (The detection and identification of these contaminants proved that the instrument was working properly.) The only other gases found were carbon dioxide and water vapor, both formed by thermal decomposition of Martian soil minerals in the GCMS ovens. Further details can be found in the paper by the molecular analysis team (Biemann et al., 1977) listed in the Bibliography.

The absence of organic matter in the Martian surface at a parts-per-billion level of detectability was the most important single biological finding of the Viking mission. With the return of the first set of GCMS data from the Chryse landing site, unambiguously negative, it became clear that if subsequent analyses continued to show no organic matter in the Martian soil, then convincing evidence for life in the soil would be unattainable, regardless of any other findings. As it turned out, other biological experiments reinforced the GCMS results.

The Microbiological Experiments

At one point in the early days of the space program, the late Albert Tyler, a well-known Caltech biologist, suggested that a good biological experiment for Mars would consist of a mousetrap and a camera. By the mid-1960s, however, the idea that Mars might harbor higher forms of life had been abandoned by almost everyone. The prevailing view was that microbial life was the most one could hope for. Even those who insisted on the theoretical possibility of higher Martian forms recognized that the chance of detecting life on the planet would be maximized by focusing on soil microorganisms. One could not imagine a planet inhabited by higher forms of life that did not also harbor microorganisms, but one could easily imagine the reverse: a planet on which the only life was microbial. Even a spacecraft from another planet bent on finding out whether our world is inhabited would do well to test the soil. The soil is a vast biological community inhabited by bacteria, yeasts, and molds. Besides being abundant, these organisms are very hardy, the last survivors in extreme environments, so that it is very unusual to find even small soil samples anywhere on the earth that are devoid of microorganisms.

Accordingly, each Viking lander carried three instruments designed to detect the metabolic activities of soil microorganisms. These experiments—all of them selected by NASA from among submitted proposals—were similar in that they exposed small samples of the Martian surface material to substances of various kinds and then monitored the fate of the substances. Heat was used to discriminate between biological and nonbiological reactions, the assumption being that a heat-stable reaction was very probably nonbiological, whereas a heat-sensitive one could be either biological or not.

Two of the experiments were frankly terrestrial in orientation, so much so that neither of them could operate under actual Martian conditions. Both employed organic compounds in aqueous solutions that were to be incubated with samples of Martian soil.

Because liquid water cannot exist on Mars, the solutions, along with their Martian soil samples, had to be heated and pressurized to values well above Martian in order to prevent them from freezing and/or boiling. These experiments seemed to be designed more for Lowellian or pre-*Mariner-4* Mars than for the real Mars, and serious questions were raised as to the wisdom of including them in the Viking payload. In the end, they remained on board, although a third aqueous experiment, also approved for the mission, was deleted in the course of building the instrument package. As it happened, the two wet experiments contributed in an entirely unexpected but very important way to our present understanding of Mars.

The Gas-Exchange Experiment. As originally conceived by its inventor, Vance Oyama, the plan of the gas-exchange experiment (GEx) was to mix a sample of Martian soil with a nutrient solution in a sealed chamber at a temperature of about 10°C. The chamber would be pressurized by an injection of gas (a mixture of helium, krypton, and carbon dioxide was used), and changes in the composition of the gas in the chamber would be monitored over a period of time. The presence of organisms in the soil would, it was thought, result in the production and disappearance of metabolic gases of various kinds, an effect that occurs in terrestrial soils. A gas chromatograph was designed to identify and measure the gases.

The selection of chemicals in the aqueous nutrient solution—it included vitamins, amino acids, purines and pyrimidines, organic acids, and inorganic salts—was strongly oriented toward terrestrial biochemistry. After discussion by the Viking biology team, which was made up of the originators of the three experiments, plus three additional members appointed by NASA[2], a small but crucial change was made in the protocol of the gas-exchange experiment. Before the soil was actually wetted with the nutrient solution, it was to be sealed in the chamber with a small volume of the solution that would be separate from the soil sample but that would saturate the

---

[2] The team members were Harold P. Klein, Joshua Lederberg, Gilbert V. Levin, Vance Oyama, Alexander Rich, and the author.

chamber with water vapor. Thus the Martian sample would be exposed to water vapor at a pressure that had not been seen on Mars for many millions of years. Gas measurements would be made in this "humid mode" before the soil came into contact with the actual solution.

This new, first phase of GEx produced a surprising result. On its exposure to water vapor, the soil released four gases: nitrogen, argon, carbon dioxide, and oxygen. The first three appeared in relatively small amounts and their presence could be explained as a displacement of absorbed gases from the samples by water vapor. In the case of oxygen, however, simple desorption could not explain the pressure increase. In the first experiment at the Chryse site, for example, oxygen pressure increased nearly 200-fold in less than two sols after the chamber was humidified. As Oyama and Berdahl (1977) pointed out at the time, so large a change implied that oxygen gas was being produced by a chemical reaction initiated by contact between water vapor and some substance in the soil. Obvious candidates for the latter role were the oxygen-rich peroxides, superoxides, and ozonides. These compounds have the general formulae M_2O_2, MO_2, and MO_3, respectively, where M is either hydrogen or a metal. They decompose readily in the presence of water to yield oxygen. Such powerful oxidizing agents in the surface of Mars would explain not only the evolution of oxygen, but also the absence of organic matter from the soil. The presence of these substances had been predicted for Mars, and they probably occur in the Martian soil in greater abundance than the Viking results indicated, as we shall see.

After seven sols of incubation in the humid mode, the first Martian sample was wetted by allowing more nutrient solution to enter the chamber. (GEx was now in its originally intended configuration.) The wet soil was incubated for 196 sols (6.7 months) to provide plenty of opportunity for Martian organisms to manifest their presence by producing or consuming metabolic gases. No such manifestation occurred. The only significant change seen during these months was a loss of oxygen as it combined with the ascorbic acid (vitamin C) of the nutrient solution.

For the second experiment, the chamber was emptied of gas and medium, dried, and reloaded with a fresh sample of surface material. The chamber was heated at 145°C for 3.5 hours, cooled, and

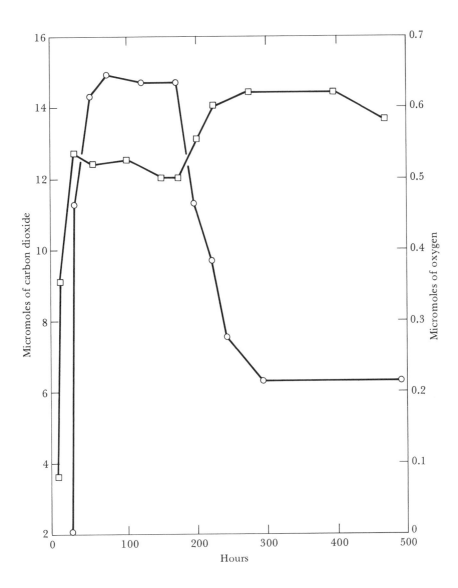

Figure 7-3 **Release of carbon dioxide (squares) and oxygen (circles) in the first gas-exchange experiment at Chryse. Hours are measured from the moment of humidification of the chamber. The curve for oxygen production is displaced 20 hours to the right in order to separate the two curves. After 175 hours of incubation in the humid mode, the sample was wetted with the medium. The loss of oxygen that followed was caused by its reaction with ascorbic acid in the medium.**

humidified again. Again oxygen was released, about half as much as in the first experiment, but in amounts sufficient to confirm that its source was not biological.

Similar results, but on a smaller scale, were obtained by Lander 2 when the gas-exchange experiment was repeated at the Utopia site. Thus, the gas-exchange experiment established that the Martian surface is chemically active, owing to a globally distributed peroxidelike material in its soil.

The Labeled-Release Experiment. Designed by Gilbert Levin, the labeled-release experiment (LR) also proceeded on the assumption that Martian organisms would generate gas from an aqueous solution of nutrients. It differed from the gas-exchange experiment in some important details, however. For one thing, its cocktail contained simpler and more universal organic compounds than the mixture of nutrients employed in GEx. Just seven substances were used, dissolved in water: formic, glycolic, and lactic acids added as their sodium or calcium salts and the amino acids glycine and alanine; both optical isomers of alanine and lactic acid were included. All of these molecules are formed in the Miller spark-discharge reaction, all have been detected in meteorites or interstellar clouds, and it is reasonable to suppose that life anywhere would recognize and metabolize one or more of these compounds. LR also differed from GEx in that the LR nutrients were labeled with radioactive carbon, and the formation of carbon-containing gas (principally CO_2) was detected by radioactivity measurements, a procedure that made the experiment very sensitive. In its combination of generality and sensitivity, LR came close to being an ideal life-sensing device for an aqueous planet.

The experiment was initiated by adding approximately 0.1 cubic centimeter of the radioactive medium to 0.5 cubic centimeters of Martian soil, along with sufficient helium to prevent the medium from boiling at the temperature of the chamber (about 10°C). The volume of medium injected was planned to wet some, but not all, of the Martian soil sample. Almost immediately after the injection, there was a surge of radioactive gas. This production of gas gradually tapered off, and eventually it amounted to nearly that expected if just one of the 17 different carbon atoms represented in the medium had been converted to radioactive CO_2. The most

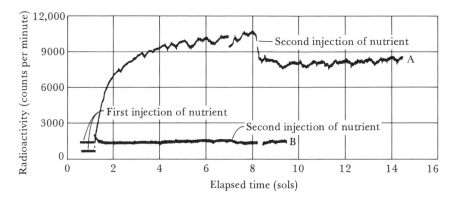

Figure 7-4   Production of radioactive carbon dioxide in the first labeled-release experiment at Chryse (curve A) and its heated control (curve B). The drop in radioactivity following the second injection of nutrient (curve A) shows that some carbon dioxide dissolved in the medium. (From "The Search for Life on Mars," by Norman H. Horowitz. Copyright © 1977 by Scientific American, Inc. All rights reserved.)

likely source of this gas was formic acid, a one-carbon compound that is easily oxidized to CO_2 by peroxides.

After the evolution of gas had nearly ceased, the nutrient solution was injected for a second time. If the radioactive gas were produced by the action of peroxides in the soil, the addition of fresh medium would cause no further gas formation since water vapor from the first injection should have decomposed the peroxides in the entire soil sample, even in portions that had not come into contact with the solution. If, however, the radioactive gas were produced by microorganisms in the soil, the addition of fresh medium should lead to the production of more gas. The result bore out the first hypothesis: no more gas was evolved. The same result was obtained with all the Martian samples tested. Logically, the inference that this nonproduction of gas corroborates the GEx findings has its precedent in the deduction made by Sherlock Holmes following a different and more famous nonoccurrence, the one called "the curious incident of the dog in the nighttime":

> "The dog did nothing in the nighttime."
> "That was the curious incident," remarked Sherlock Holmes.

The next step in the LR sequence was to repeat the test with a heat-treated soil sample. In the GEx experiment, about half the oxygen-generating capacity was lost on heating at 145°C for 3.5 hours. In the LR tests, however, all activity was lost from Martian soil heated to 160°C for three hours. The difference in length of time and temperature is unimportant. A difference in procedure, however, may have been significant, as Oyama later pointed out. The GEx chamber was kept open during the heating, with a stream of helium running through it, whereas the LR chamber was closed. The GCMS results showed that up to 1 percent of their weight in water is evolved from Martian samples heated for 30 seconds at 500°C, and some water is also formed at 200°C. This moisture doubtless originates in hydrated minerals, not free water. No GCMS tests were performed at 160°C, but sufficient water may well have been generated during the long exposure of Martian soil to this temperature in the LR chamber to have destroyed the oxidizing agent responsible for the production of CO_2. Another possibility is that heat-stable and heat-labile peroxides exist on Mars and that the peroxide responsible for the oxidation observed in the LR experiment belonged to the latter class.

The Pyrolytic Release Experiment.   One of the few statements about Martian life that could be made with confidence before the Viking mission was that if it existed, it was living under Martian, not terrestrial, conditions. Accordingly, the pyrolytic release (or PR) experiment, also called the carbon assimilation experiment, was designed by me and my collaborators, George Hobby and Jerry Hubbard, to perform a biological test on Martian soil under essentially Martian conditions.

The plan of the experiment was to expose a Martian surface sample to a Martian atmosphere that had been enriched with a small quantity of radioactively labeled CO and CO_2—these two gases are present, unlabeled, in the Martian atmosphere in amounts of 0.1 and 95 percent, respectively—and then to measure the incorporation of radioactive carbon atoms into organic matter in the sample. The soil would be incubated under Martian pressure, temperature, and atmospheric composition, in Martian sunlight, for 120 hours. The radioactive atmosphere would then be removed from the chamber and the soil sample heated to 625°C under a stream of helium in order to pyrolyze any organic substances present and convert them into volatile fragments. The fragments would be swept out of the chamber into a column packed with a material—a form of diatomaceous earth—that absorbs organic molecules, but not CO or CO_2. Once the organic molecules had been separated from the unreacted radioactive gases, the column would be heated to $640°C$ to release them and allow them to be oxidized to CO_2 by copper oxide in the column. Finally, the radioactivity of the CO_2 would be measured.

On Mars, the experiment operated as planned in all but two respects. First, owing to heat sources within the landers, the temperatures of the test cells remained above Martian ground temperatures at both landing sites. The cell temperatures ranged from 8 to $26°C$, but the ground outside remained below $0°C$ throughout the mission. Since Martian equatorial temperatures reach $25°C$, however, the chamber temperatures were not altogether un-Martian.

Second, the source of illumination employed in the experiment was not the Martian sun—design difficulties had prevented us from using it—but a xenon lamp that simulated sunlight at the surface of Mars but from which wavelengths shorter than 320 nanometers were filtered out. The purpose of the light was to provide energy for photosynthesizing organisms. Laboratory experiments had shown that a nonbiological synthesis of simple organic compounds from CO and water vapor takes place on mineral surfaces irradiated with ultraviolet below 300 nanometers, however, and we decided to remove these wavelengths in order to avoid any possibility of confusing biological and nonbiological sources of organic matter. Although this portion of the spectrum is part of the normal Martian environment, we justified its removal on the grounds that it is so destructive to complex organic molecules that Martian organisms would have to avoid it or filter it out themselves in order to survive.

Laboratory tests had shown that the experiment did not depend on photosynthesis in soil samples for its operation. The fixation of CO and CO_2 into organic matter in living cells also occurs by dark processes, and laboratory tests of the PR experiment had detected fixation of this sort as well as photosynthesis.

A total of nine PR experiments were completed on Mars, six at the Chryse site and three at Utopia. The very first test (performed at Chryse and labeled C1 in the accompanying diagram) gave a positive result. The amount of carbon fixed was small in comparison to that found when using terrestrial soil samples, but it was far above the background level established in preflight laboratory tests on sterile samples. In view of the efforts that had been made to remove nonbiological interference from the experiment, the return of even a feeble signal from Mars was surprising. It was therefore decided to make the next experiment (C2) a control: a second Martian soil sample was heated to 175°C for three hours before incubating it with the radioactive gases. The amount of carbon fixed dropped 88 percent. It appeared that we had found a heat-sensitive synthesis of organic matter on Mars, but the fact that 12 percent of the reaction survived the high temperature argued against a biological interpretation.

The next two experiments (C3 and C4) tried but failed to repeat C1. Both experiments gave weakly positive results, judged by the criteria established in the preflight tests, but neither of them came close to C1 in the quantity of carbon fixed. C5 was another heated control, performed to test the thermostability of the small reactions seen in C3 and C4. This time, the soil sample was brought to 120°C for about two minutes, then the temperature was allowed to fall to 90°C, where it remained for nearly two hours. This treatment produced no change in the reaction, again suggesting that it was not biological in origin. C6, the last experiment at the Chryse

Figure 7-5 Results of the pyrolytic-release experiments performed on Mars at Chryse (labeled C1 through C6), Utopia (U1 through U3), and in the laboratory using maghemite (M3 through M6) under simulated Martian conditions. Peak 1 measures the amount of radioactive gas that was adsorbed onto the sample without undergoing further reaction. Peak 2 measures the fraction of adsorbed gas that was converted into organic matter. Points falling above the sloping line near the bottom of the graph are significantly higher than the background established in preflight tests with sterilized soils or with no soils. The maghemite samples were dried and degassed, held for five days under Martian conditions of atmospheric composition, pressure, and ultraviolet flux, and then tested in a duplicate of the Viking pyrolytic release instrument. (Further details appear in the paper by Hubbard listed in the Bibliography.)

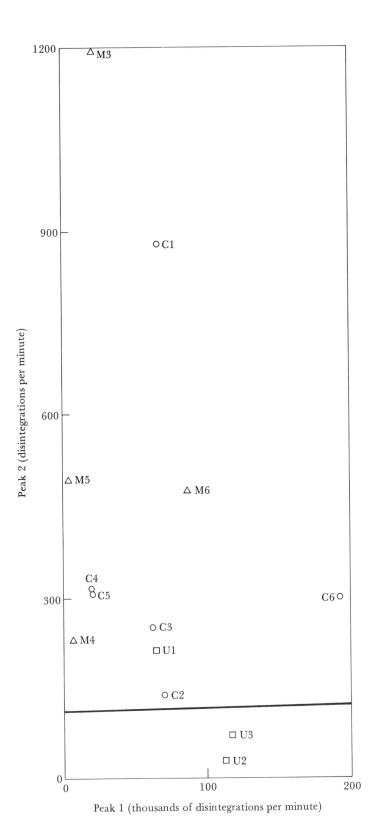

site, tested the effect of water vapor on the reaction. No effect was found.

Of the three experiments performed at the Utopia site, U1 resembled C2 through C6 in its weak positive response. U2 and U3 were negative by the preflight criteria. Further tests at Utopia were precluded by the development of a leak in the instrument.

Although the positive signals obtained in the PR experiments are still not completely understood, the chance that their source was biological seems negligible. The reasons for this conclusion are these:

1. Our inability to duplicate the promising result of C1 suggests that the high C1 reading was an anomaly, caused perhaps by an instrument malfunction. If so, the 88 percent loss of activity found in C2 is spuriously high, and the reaction is more stable to heat than it appears to be. A thermostable, nonbiological reaction is clearly indicated in C5.

2. Water should be the limiting factor for life on Mars (the reasons for this were discussed in Chapter 6), yet injection of an amount of water vapor calculated to bring the test-cell humidity close to saturation had either no effect on the response or a deleterious one. (Water was injected in experiments C5, C6, and U2. For details, see the papers by Horowitz et al. in the Bibliography.)

3. Although the data are few, it appears that the reaction is about the same in the dark as the light. (U1 and U3 were carried out in the dark; all others were in the light.) Soil samples from the surface of the earth typically fix much more carbon in the light, owing to the presence of photosynthesizing organisms.

4. Laboratory simulations performed since the Viking mission have shown that except for the questionable sensitivity to high temperature, all of the characteristics mentioned above are duplicated in a nonbiological reaction between the radioactive gas mixture and any of several iron-rich minerals. One such mineral is maghemite, or $\gamma-Fe_2O_3$, a magnetic form of iron oxide that is of minor significance on the earth but that Viking evidence suggests may be widespread on Mars.

The results now available thus suggest that one or more iron-containing minerals in the Martian surface reacting, probably, with the CO of the gas mixture explains the fixation of carbon detected in the PR experiment. The surface of Mars contains 13 percent iron. Although specialists are still discussing the question of just what minerals are present, it is likely that an iron-catalyzed reaction was the source of the signals detected by the PR instrument. The nature of the product, whether organic—that is, containing carbon atoms bonded to hydrogen—or inorganic, is unknown. If the former, the amount of organic synthesis implied by the PR data was close to the detection limit of the GCMS, assuming the carbon was fixed into a single compound. If more than one compound was made, the GCMS could not have detected them. In any case, the GCMS and PR results do not conflict.

The problem of reconciling the PR findings with the evidence for destructive peroxy compounds in the Martian surface is another matter. If peroxy compounds are homogeneously distributed in the soil, synthesis of organic compounds is not expected. In those PR experiments where water vapor was injected with the radioactive gases, however, there was no observable increase in the amount of carbon fixed, and this suggests that the distribution is not homogeneous. It would follow, then, that the soil grains that were active in fixing carbon in the PR experiment were not those to which peroxy compounds attach.

Generality of the Findings

The two Viking landing sites were very similar in their surface chemistry, despite their climatic differences and the great distance between them. We now know that this similarity is a consequence of processes that are planetwide in their operation; the findings made at Chryse and Utopia are probably typical of the entire planet. One such process consists of planetary windstorms that distribute fine surface materials around the globe. Another, especially important because of its biological implications, consists of the splitting of water molecules in the lower atmosphere of Mars by short-wavelength solar ultraviolet rays. The photolytic products of this irradiation, H and OH, are very reactive species, and their subsequent fate has been the subject of illuminating theoretical studies by Donald Hunten and other aeronomists.

OH is a powerful oxidizing agent whose production near the

Martian surface explains the absence of organic matter. Its presence, incidentally, explains why Mars is red—it is covered with oxides of iron. Further, it explains why the Martian atmosphere does not consist of CO and O_2. This is the mixture formed when CO_2 is irradiated with solar ultraviolet, but in the presence of OH, the CO is reoxidized. Finally, reactions involving OH readily lead to the formation of peroxy compounds such as H_2O_2 and HO_2. Even before the Viking mission, Hunten had predicted that these compounds would migrate into the Martian surface from the atmosphere. These are, of course, just the substances needed to account for the GEx and LR findings.

The aeronomic predictions came too late to affect the design of the Martian biology experiments. Indeed, we did not even know about them until after the mission. The fact that the Viking instruments nevertheless confirmed them involved considerable good luck, not only in the eleventh-hour change of the GEx protocol, but in another way, as well. In their descent to Mars, both spacecraft moved across ground that would later be sampled with their retroengines firing. The exhaust contained 0.5 percent water vapor, and it seems inevitable that some destruction of peroxy compounds occurred. In addition, one-third of the exhaust was composed of ammonia, a combustible gas that could also have reacted with these substances in the presence of catalysts in the soil. It is thus likely that the soil samples that were finally tested contained only a fraction of their original reactive molecules. That any remained was sheer good luck.

Oxidation by peroxides is not the only way that organic matter can be destroyed on Mars. Experiments by Pang and his colleagues have shown that in the presence of ultraviolet light and titanium oxide (the Martian surface contains 0.5 percent titanium), atmospheric oxygen promotes the rapid oxidation of organic surface matter. Like OH oxidation, this process also operates globally.

With at least two planetwide mechanisms acting to destroy organic matter on the surface of Mars, there is little doubt that the Viking findings represent the entire planet.

CHAPTER EIGHT

Summary and Conclusions: Life in the Solar System

We shall not cease from exploration
And the end of all our exploring
Will be to arrive where we started
And know the place for the first time.
T. S. Eliot, Little Gidding

The belief that the planets are abodes of life goes back at least 300 years to a time before much was known about either life or the planets. A natural but unwarranted extension of the Copernican revolution, this notion was accepted by thinkers of the seventeenth and eighteenth centuries on the basis of general philosophical principles, not evidence. In time, as science became more sophisticated, the existence of life on other planets came to be regarded not as a self-evident truth but as a question, or hypothesis, to be elaborated deductively and tested against observation. The fulfillment of this program waited on two advances that came about only in our time: a deeper understanding of the nature and origin of living matter, and the invention of new methods for studying the planets that transcended the limitations of earth-based telescopes. The new methods took the form of interplanetary spacecraft and their increasingly sophisticated communication technologies.

Biologists in our century have shown that life is a chemical phenomenon that differs from other phenomena in exhibiting genetic properties. In all living systems known to us, the agents of these properties are nucleic acids and proteins. The similarities of nucleic acids, proteins, and genetic operating mechanisms among the most diverse species leave little doubt that all organisms presently living on the earth are related through an evolutionary web that also connects them with extinct species that existed in past ages. Such evolution is the natural and inevitable result of the operation of genetic systems. We are led to conclude that despite their endless variety, the living things on our planet are members of one family. There is in fact just one form of life on the earth, and the origin of life may have happened only once.

Carbon is the characteristic element of terrestrial biochemistry. The chemical properties of carbon make it especially suitable for the construction of large, information-rich molecules of the kind needed for building genetic systems with virtually unlimited evolutionary capabilities. The abundance of carbon in the cosmos is also very great, and a variety of evidences—laboratory experiments, analysis of meteorites, interstellar spectroscopy—show that the formation of organic compounds like those found in living matter occurs readily and on a large scale in the universe. It is therefore probable that if life exists elsewhere in the universe, it also is based on carbon chemistry.

For a carbon-based biochemistry to manifest itself, certain conditions of planetary temperature and pressure are required, along with a suitable energy source, a planetary atmosphere, and a solvent. Although water is the solvent of terrestrial biochemistry, it is conceivable, but not certain, that other solvents could function in extraterrestrial biochemistries.

In the light of conditions as they actually exist on the known planets, these minimal environmental requirements for life are found to be quite restrictive, and habitable planets are therefore quite rare. Thanks to substantial progress in our knowledge of the planets, it had become clear by 1975 that only Mars could be considered as a remotely possible habitat for extraterrestrial life in our solar system. The historic Viking mission launched in that year, the last in a series of important spaceflights to Mars, climaxed the search for life on the planets. It also brought to a close one of the strangest chapters in the annals of modern science, in the course of which a fictional Mars had been created. The unraveling of the Martian myth, which began in 1963, told us as much about people as it did about Mars. It also demonstrated the power of science to discover and correct its own errors.

Viking found no life on Mars, and, just as important, it found why there can be no life. Mars lacks that extraordinary feature that dominates the environment of our own planet, oceans of liquid water in full view of the sun; indeed, it is devoid of any liquid water whatsoever. It is also suffused with short-wavelength ultraviolet radiation. Each of these circumstances alone would probably suffice to ensure its sterility, but in combination they have led to the development of a highly oxidizing surface environment that is incompatible with the existence of organic molecules on the planet. Mars is not only devoid of life, but of organic matter as well.

For some, Mars will always be inhabited, no matter what the data say. Occasionally one hears the opinion that somewhere on the planet there may exist a wet, warm place—a Martian Garden of Eden—where Martian life forms are thriving. Or, alternatively, that the Viking instruments did in fact find life—that the Viking data can be interpreted to mean that there are organisms living in the soil at a population density below the GCMS detection limit.

These contradictory views—one assuming that Martian life is like our own in its need for water, the other that it is not—are daydreams. The Garden of Eden would identify itself in photographs by a permanent water cloud above it and, probably, by snow on the ground. These signs have not been seen, and it is extremely unlikely that any such place exists on Mars. The Utopia landing site, where frost covers the ground for long periods of each year, is very watery by Martian standards, so it is not correct to say that the Viking mission sampled only the most desiccated areas. The second idea, that microorganisms are even now living in the Martian soil, is just another form of the blue unicorn theory. According to this theory, a blue unicorn is living in a cave on the moon, an assertion that is impossible to disprove because the unicorn is endowed by its inventor with whatever attributes are found necessary to allow it to survive on the moon. In the case of organisms on Mars, these would include the ability to live without water or any other solvent and immunity from the processes that destroy all other forms of organic matter on the planet.

The failure to find life on Mars was a disappointment, but it was also a revelation. Since Mars offered by far the most promising habitat for extraterrestrial life in the solar system, it is now virtually certain that the earth is the only life-bearing planet in our region of the galaxy. We have awakened from a dream. We are alone, we and the other species, actually our relatives, with whom we share the earth. If the explorations of the solar system in our time bring home to us a realization of the uniqueness of our small planet and thereby increase our resolve to avoid self-destruction, they will have contributed more than just science to the human future.

Glossary

Adsorption The binding of gas molecules or molecules in solution to solid surfaces by specific physical or chemical forces.

Aldehyde An organic compound of the general structure $RC\!\!\overset{\displaystyle H}{\underset{\displaystyle |}{=}}\!\!O$, where R is either hydrogen (giving formaldehyde) or an organic radical. The symbol $=$ represents a double bond (see *Chemical Bond*).

Amino acid The basic subunit of proteins, of the general formula $RCH(NH_2)COOH$, where R is any one of twenty different radicals.

Catalyst A substance that accelerates a chemical reaction without being consumed or changed thereby.

Chemical bond The force holding atoms together in molecules. The most common chemical bond consists of a pair of electrons shared by two atoms. A double bond consists of two pairs, and a triple bond of three pairs of shared electrons. Hydrogen bonds, important in the structure of water, DNA, and proteins consist of a hydrogen ion shared by two electronegative atoms such as oxygen or nitrogen.

Dielectric constant The property of matter of interacting with electric charges in such a way as to reduce the force between opposite charges. The dielectric constant is important in explaining the solvent properties of liquids. Water has one of the highest dielectric constants of any liquid.

Electrolyte A substance that, when dissolved in water, dissociates into positively and negatively charged ions.

Electrostriction The change in volume and freedom of movement of solvent molecules associated with their binding by electric fields produced in many chemical reactions.

Enzyme A protein that functions as a biological catalyst.

Free energy A measure of the amount of work (chemical, mechanical, or electrical) associated with chemical reactions taking place at constant pressure and temperature. Reactions that release free energy—the oxidation of sugar, for example—occur spontaneously, and they can serve as energy sources for the performance of useful work. Reactions that consume free energy—photosynthesis, for example—cannot occur without an input of energy, and these reactions store free energy. Reactions for which the free energy change is zero are said to be at equilibrium.

Genetic system The chemical substances and physical arrangements that underlie the properties of self-replication and mutation.

Greenhouse effect The heating of the atmosphere caused by opacity of the atmosphere to infrared radiation. The latter radiation results when the ground absorbs solar wavelengths (mostly visible) to which the atmosphere is transparent.

Hoarfrost Minute ice crystals formed by condensation of a gas onto a cold surface.

Hydrocarbon A chemical compound made up of the elements carbon and hydrogen only.

Hydrogen bond See *Chemical bond.*

Ion An atom or group of atoms carrying an electric charge due to the loss or gain of one or more electrons or protons.

Isotope One of the alternative forms of a chemical element that differ only in atomic mass. Isotopes of the same element are alike in chemical properties, but some are unstable and disintegrate with the production of radioactivity.

Katabatic wind A downslope wind. Such winds are heated by compression and are characterized by low relative humidities.

Mantle The layer of the earth that lies between the crust at the surface and the core at the center.

Micromole A millionth of a mole. A mole is the weight of a substance, in grams, equal to its molecular weight. A mole contains 6.02×10^{23} molecules of the substance.

Monomer A molecule (such as a nucleotide or amino acid) from which polymers (such as nucleic acids and proteins) are formed, usually by linear catenation.

Mutation A random change in gene structure that is perpetuated by self-duplication.

Nucleic acid A linear polymer formed from nucleotides by combining the phosphoric acid of one nucleotide with the sugar of the next. One kind of nucleic acid, DNA, forms the genes. Another kind, RNA, is important in protein synthesis.

Nucleotide The monomeric subunit of nucleic acids, of the general structure base-sugar–phosphoric acid.

Optical isomer A chemical compound that, in solution, rotates the plane of polarized light either clockwise or counterclockwise. For every such compound there is another, the mirror image of the first, that rotates polarized light in the opposite direction.

Oxidation The addition of oxygen to an element or compound, or the subtraction of hydrogen or electrons.

Photolysis The splitting of molecules caused by absorption of (usually) ultraviolet radiation.

Photosynthesis The process by which green plants, algae, and some bacteria utilize sunlight for synthesizing organic material from carbon dioxide.

Polar molecule A molecule in which the center of positive charge does not coincide with the center of negative charge, the result being the existence of positive and negative poles. Most of the molecules of living cells are polar, as are such solvents as water, ammonia, and the alcohols. Polar solvents have high dielectric constants (q.v.).

Polypeptide A linear polymer formed by the end-to-end linking of amino acids.

Polymer See *Monomer.*

Protein A molecule composed of one or more polypeptides (q.v.). Proteins, in the form of enzymes (q.v.), have a central role in practically all chemical reactions of living cells. In other forms, they fulfill many other biological roles, as well—for example, they make up muscle fibers.

Radical An atom or group of atoms possessing one or more unpaired electrons. Free radicals, that is, radicals not combined into molecules—for example, H and OH formed by the photolysis of water—are usually very reactive.

Reduction The addition of hydrogen or electrons to, or the subtraction of oxygen from, an element or compound.

Relative humidity The amount of water vapor in air expressed as a percentage of the amount that would be present in saturated air at the same temperature.

Spectrogram A photograph of a spectrum.

Sublimation Evaporation of a solid without melting.

Sugar A carbohydrate usually containing 12 or fewer carbon atoms and having the general formula $C_m(H_2O)_n$. Ordinary table sugar, or sucrose, is $C_{12}H_{22}O_{11}$.

Triple point The temperature at which the liquid, solid, and vapor phases of a substance are in equilibrium. The triple point of pure water is at $0.0099°C$.

Vapor pressure The pressure exerted by a vapor in equilibrium with its liquid or solid phase.

Bibliography

Chapter One

Dayhoff, M. O. 1972. *Atlas of Protein Sequence and Structure,* vol. 5. National Biomedical Research Foundation, Washington, D.C.

Edsall, J. T., and J. Wyman. 1958. *Biophysical Chemistry.* Academic, New York. A classic work with an excellent introduction to the importance of carbon and water in living systems.

Horowitz, N. H. 1979. Genetics and the synthesis of proteins. In P. R. Srinivasan, J. S. Fruton, and J. T. Edsall, eds., *The Origins of Modern Biochemistry. Ann. N.Y. Acad. Sci.* **325:**253–266.

Muller, H. J. 1929. The gene as the basis of life. *Proc. Internatl. Cong. Plant Sci.* **1:**897–921. Reprinted in part in H. J. Muller (1962), *Studies in Genetics.* Indiana University Press, Bloomington.

Plapp, F. W. 1976. Biochemical genetics of insecticide resistance. *Ann. Rev. Entomol.* **21:**179–197.

Stryer, L. 1981. *Biochemistry,* 2d ed. W. H. Freeman and Co., New York. An introductory textbook, especially strong on the chemistry of proteins and nucleic acids.

Chapter Two

Aristotle. *Historia Animalium.* Translated by D'Arcy Wentworth Thompson. Oxford, 1910.

Arrhenius, Svante. 1908. *Worlds in the Making.* Harper & Bros., New York.

Crick, Francis. 1981. *Life Itself—Its Origin and Nature.* Simon and Schuster, New York.

Crick, F. H. C., and L. E. Orgel. 1973. Directed panspermia. *Icarus* **19:**341–346.

Della Porta, Giambattista. 1558. *Natural Magick.* English edition of 1658, © 1957 by Basic Books, New York.

Fisher, R. B. 1977. *Joseph Lister.* Stein and Day, New York.

Hart, M. H., and B. Zuckerman. 1982. *Extraterrestrials—Where Are They?* Pergamon, New York.

Helmholtz, H. 1884. *Popular Scientific Lectures.* Longmans, London.

Hoyle, F., and C. Wickramasinghe. 1979. On the nature of interstellar grains. *Astrophys. Space Sci.* **66**:77–90.

Pasteur, Louis. 1922. *Oeuvres,* collected by Pasteur Vallery-Radot. Masson et Cie., Paris.

Redi, Francesco. 1688. *Experiments on the Generation of Insects.* Translated by Mab Bigelow, 1909. Open Court, Chicago.

Tyndall, John. 1897. *Fragments of Science,* vol. 2. D. Appleton & Co., New York.

Chapter Three

Anders, E., R. Hayatsu, and M. H. Studier. 1973. Organic compounds in meteorites. *Science* **182**:781–790.

Bernal, J. D. 1967. *The Origin of Life.* World, Cleveland. An appendix contains an English translation of Oparin's 1924 paper and a reprint of Haldane's 1929 essay referenced below.

Cairns-Smith, A. G. 1975. A case for alien ancestry. *Proc. Roy. Soc. Lond. B.* **189**:249–274.

Cameron, A. G. W. 1970. Abundances of the elements in the solar system. *Space Sci. Reviews* **15**:121–146.

Cronin, J. R., W. E. Gandy, and S. Pizzarello. 1980. Amino acids in the Murchison meteorite. In P. E. Hare, T. C. Hoering, and K. King, eds., *The Biogeochemistry of Amino Acids.* Wiley, New York, p. 153.

Cronin, J. R., and S. Pizzarello. 1983. Amino acids in meteorites. *Adv. Space Res.* **3**:5–18.

Haldane, J. B. S. 1929. The origin of life. *Rationalist Annual.* Reprinted in Bernal, referenced above.

Herbst, E., and W. Klemperer. 1976. The formation of interstellar molecules. *Physics Today* **1976**:32–39.

Inoue, T., and L. E. Orgel. 1983. A nonenzymatic RNA polymerase model. *Science* **219**:859–862.

Mason, B. 1966. *Principles of Geochemistry,* 3d ed. Wiley, New York.

Miller, S. L., and L. E. Orgel. 1974. *The Origins of Life on the Earth.* Prentice-Hall, Englewood Cliffs, N.J.

Miller, S. L. 1982. Prebiotic synthesis of organic compounds. In H. D. Holland and M. Schidlowski, eds., *Mineral Deposits and the Evolution of the Biosphere.* Springer-Verlag, New York.

Oparin, A. I. 1938. *The Origin of Life.* Macmillan, New York.

Rabinowitch, E. I. 1945. *Photosynthesis,* vol. 1. Interscience, New York, pp. 85–88. A discussion of the experiments of Baly cited by Haldane in support of his theory of the origin of life.

Rank, D. M., C. H. Townes, and W. J. Welch. 1971. Interstellar molecules and dense clouds. *Science* **174**:1083–1101.

Urey, H. C. 1952. On the early chemical history of the earth and the origin of life. *Proc. Natl. Acad. Sci.* **38**:351–363.

Urey, H. C. 1952. *The Planets.* Yale University Press, New Haven.

Walker, J. C. G. 1977. *Evolution of the Atmosphere.* Macmillan, New York.

Wood, J. A. 1979. *The Solar System.* Prentice-Hall, Englewood Cliffs, N. J.

Chapter Four

Brock, T. D. 1967. Life at high temperatures. *Science* **158**:1012–1019.

Hamann, S. D. 1957. *Physico-Chemical Effects of Pressure.* Butterworths, London.

Henderson, L. J. 1913. *The Fitness of the Environment.* Macmillan, New York.

Horowitz, N. H. 1976. The search for life in the solar system. *Accts. Chem. Res.* **9**:1–7.

Huygens, C. 1698. *Cosmotheoros, or New Conjectures Concerning the Planetary Worlds.* Excerpted in D. Goldsmith, ed., *The Quest for Extraterrestrial Life.* University Science Books, Mill Valley, Calif., 1980.

Ingersoll, A. P. 1982. Jupiter and Saturn. In J. K. Beatty, B. O'Leary, and A. Chaikin, eds., *The New Solar System,* 2d ed. Cambridge University Press, pp. 117–128. An interesting and authoritative discussion of the physics and chemistry of the two largest planets, for the general reader. See also the articles by Johnson and by Pollack, cited below, from the same collection.

Jannasch, H. W., and C. O. Wirsen. 1979. Chemosynthetic primary production at East Pacific sea floor spreading centers. *BioScience* **29**:592–598. Describes experiments on the metabolism of microorganisms from the geothermal vents on the floor of the Pacific Ocean.

Jannasch, H. W., and M. J. Mottl. 1985. Geomicrobiology of deep-sea hydrothermal vents. *Science* **229**:717–725. A recent review with new results.

Johnson, F. H., H. Eyring, and B. J. Stover. 1974. *The Theory of Rate Processes in Biology and Medicine.* Wiley, New York. Describes pressure-temperature relations in biological systems.

Johnson, T. V. 1982. The Galilean satellites. In J. K. Beatty, B. O'Leary, and A. Chaikin, eds., *The New Solar System,* 2d ed. Cambridge University Press, pp. 143–160.

Kant, I. 1755. *Allgemeine Naturgeschichte und Theorie des Himmels.* Ostwald's Klassiker der Exakten Wissenschaften, No. 12. W. Englemann, Leipzig.

Lunine, J. I., D. J. Stevenson, and Y. L. Yung. 1983. Ethane ocean on Titan. *Science* **222:**1229–1230.

Pauling, L. 1970. *General Chemistry.* W. H. Freeman and Co., New York.

Pollack, J. B. 1982. Atmospheres of the terrestrial planets. In J. K. Beatty, B. O'Leary, and A. Chaikin, eds., *The New Solar System,* 2d ed. Cambridge University Press, pp. 57–70.

Pollack, J. B. 1982. Titan. *Ibid.,* pp. 161–166.

Prinn, R. G., and T. Owen. 1976. Chemistry and spectroscopy of the Jovian atmosphere. In T. Gehrels, ed., *Jupiter.* University of Arizona Press, Tucson, pp. 319–371. Includes a critique of the idea that Jupiter offers a suitable habitat for life.

Sagan, C. 1971. The solar system beyond Mars: An exobiological survey. *Space Sci. Rev.* **11:**827–866.

Stetter, K. O. 1982. Ultrathin mycelia-forming organisms from submarine volcanic areas having an optimum growth temperature of 105°C. *Nature* **300:**258–260.

Stone, E. C., and E. D. Miner. 1981. Voyager 1 encounter with the Saturnian system. *Science* **212:**159–162.

Walker, J. C. G. 1977. *Evolution of the Atmosphere.* Macmillian, New York.

Wolfe, R. S. 1971. Microbial formation of methane. *Advances in Microbial Physiology* **6:**107–146.

Young, A. T. 1973. Are the clouds of Venus sulfuric acid? *Icarus* **18:**564–583.

Chapter Five

Carr, M. H. 1981. *The Surface of Mars.* Yale University Press, New Haven. A well-illustrated synthesis of the geological findings of the Viking mission, by a member of the Viking team.

Cohen, M. R., and E. Nagel. 1934. *An Introduction to Logic and Scientific Method.* Harcourt, Brace, New York.

De Vaucouleurs, G. 1954. *Physics of the Planet Mars.* Faber & Faber, Ltd., London.

Dollfus, A. 1961. Polarization studies of planets. In G. P. Kuiper and B. M. Middlehurst, eds., *Planets and Satellites.* University of Chicago Press, pp. 343–399.

Glasstone, S. 1968. *The Book of Mars.* National Aeronautics and Space Administration, Washington, D.C. A historical summary of knowledge about Mars up to 1968.

Herr, K. C., and G. C. Pimentel. 1969. Infrared absorptions near three microns recorded over the polar cap of Mars. *Science* **166**:496–499.

Hoyt, W. G. 1976. *Lowell and Mars.* University of Arizona Press, Tucson.

Kaplan, L. D., G. Münch, and H. Spinrad. 1964. An analysis of the spectrum of Mars. *Astrophys. J.* **139**:1–15.

Kellogg, W. W., and C. Sagan. 1961. *The Atmospheres of Mars and Venus.* Publication 944, National Academy of Sciences-National Research Council, Washington, D.C. This is the pre-1963 panel report quoted in this chapter.

Kliore, K., D. L. Cain, G. S. Levy, V. R. Eshleman, G. Fjeldbo, and F. D. Drake. 1965. Occultation experiment: Results of the first direct measurement of Mars' atmosphere and ionosphere. *Science* **149**:1243–1248.

Kuiper, G. P. 1952. *The Atmospheres of the Earth and Planets,* 2d ed. University of Chicago Press.

Leighton, R. L., and B. C. Murray. 1966. Behavior of carbon dioxide and other volatiles on Mars. *Science* **153**:136–144.

Lowell, P. 1898. Conclusions with regard to the present condition of the planet. *Ann. Lowell Observatory* **1**:247–250.

Lowell, P. 1906. *Mars and Its Canals.* Macmillan, New York.

Lowell, P. 1908. *Mars as the Abode of Life.* Macmillan, New York.

Masursky, H. 1982. Mars. In J. K. Beatty, B. O'Leary, and A. Chaikin, eds., *The New Solar System,* 2d ed. Cambridge University Press, pp. 83–92.

Michaux, C. M., and R. L. Newburn. 1972. *Mars Scientific Model.* JPL Document No. 606–1, Jet Propulsion Laboratory, California Institute of Technology, Pasadena.

Miller, S. L., and W. D. Smythe. 1970. Carbon dioxide clathrate in the Martian ice cap. *Science* **170**:531–533.

Murray, B., M. C. Malin, and R. Greeley. 1981. *Earthlike Planets.* W. H. Freeman and Co., New York. The geology of the inner planets and the moon compared. Comprehensible to beginners.

Mutch, T. A., R. E. Arvidson, J. W. Head III, K. L. Jones, and R. S. Saunders. 1976. *The Geology of Mars.* Princeton University Press. Martian geology as it looked after the *Mariner-9* mission.

Neugebauer, G., G. Münch, S. C. Chase, Jr., and E. Miner. 1971. Mariner 1969 infrared radiometer results: Temperature and thermal properties of the Martian surface. *Astron. J.* **76**:719–728.

Rea, D. G., B. T. O'Leary, and W. M. Sinton. 1965. Mars: The origin of the 3.58- and 3.69-micron minima in the infrared spectra. *Science* **147**:1286–1288.

Shirk, J. S., W. A. Haseltine, and G. C. Pimentel. 1965. Sinton bands: Evidence for deuterated water on Mars. *Science* **147**:48–49.

Sinton, W. M. 1957. Spectroscopic evidence for vegetation on Mars. *Astrophys. J.* **126**:231–239.

Sinton, W. M. 1959. Further evidence of vegetation on Mars. *Science* **130**:1234–1237.

Chapter Six

Barker, E. S., R. A. Schorn, A. Woszczyk, R. G. Tull, and S. J. Little. 1970. Mars: Detection of atmospheric water vapor during the southern hemisphere spring and summer seasons. *Science* **170**:1308–1310. A pre-Viking study.

Benoit, R. E., and C. L. Hall. 1970. The microbiology of some dry valley soils of Victoria Land, Antarctica. In M. W. Holdgate, ed., *Antarctic Ecology*, vol. 2. Academic Press, New York, pp. 697–701.

Brown, A. D. 1976. Microbial water stress. *Bacteriol. Rev.* **40**:803–846.

Cameron, R. E. 1972. Microbial and ecologic investigations in Victoria Valley, southern Victoria Land, Antarctica. In G. A. Llano, ed., *Antarctic Terrestrial Biology*, Antarctic Research Series **20**:195–260. American Geophysical Union, Washington, D.C.

Cameron, R. E., J. King, and C. N. David. 1970. Microbiology, ecology, and microclimatology of soil sites in dry valleys of southern Victoria Land, Antarctica. In M. W. Holdgate, ed., *Antarctic Ecology*, vol. 2. Academic Press, New York, pp. 702–716.

Clark, B. C., and D. C. Van Hart. 1981. The salts of Mars. *Icarus* **45**:370–378. A discussion of the probable assemblage of salts to be found on the surface of Mars, based on Viking data. The possibility of brine pools is examined.

Davies, D. W. 1979. The relative humidity of Mars' atmosphere. *J. Geophys. Res.* **84**:8335–8340.

Edney, E. B. 1957. *The Water Relations of Terrestrial Arthropods*. Cambridge University Press.

Farmer, C. B. 1976. Liquid water on Mars. *Icarus* **28**:279–289.

Farmer, C. B., D. W. Davies, and D. D. LaPorte. 1976. Mars: Northern summer ice cap—water vapor observations from Viking 2. *Science* **194**:1339–1340.

Farmer, C. B., and P. E. Doms. 1979. Global seasonal variation of water vapor on Mars and the implications for permafrost. *J. Geophys. Res.* **84**:2881–2888.

Friedmann, E. I. 1982. Endolithic microorganisms in the Antarctic cold desert. *Science* **215**:1045–1053.

Horowitz, N. H., R. E. Cameron, and J. S. Hubbard. 1972. Microbiology of the dry valleys of Antarctica. *Science* **176**:242–245.

Horowitz, N. H., R. P. Sharp, and R. W. Davies. 1967. Planetary contamination I: The problem and the agreements. *Science* **155**:1501–1505. A critique of the plan to heat-sterilize Mars landers. See Murray et al. for part II.

Ingersoll, A. P. 1970. Mars: Occurrence of liquid water. *Science* **168**:972–973.

Kieffer, H. H. 1979. Mars south polar spring and summer temperatures: residual CO_2 frost. *J. Geophys. Res.* **84**:8263–8288.

Kieffer, H. H., S. C. Chase, T. Z. Martin, E. D. Miner, and F. D. Palluconi. 1976. Martian north polar summer temperatures: Dirty water ice. *Science* **194**:1341–1343.

Lange, O. L., E.-D. Schulze, L. Kappen, U. Buschbom, and M. Evenari. 1975. Adaptations of desert lichens to drought and extreme temperatures. In N. F. Hadley, ed., *Environmental Physiology of Desert Organisms.* Dowden, Hutchinson & Ross, Stroudsburg, Pa.

Machin, J., M. J. O'Donnell, and P. A. Coutchie. 1982. Mechanisms of water vapor absorption in insects. *J. Exp. Zool.* **222**:309–320. A good review.

Mazur, P. 1970. Cryobiology: The freezing of biological systems. *Science* **168**:939–949. A review and extensive bibliography dealing with the effects of low temperatures on cells.

Meyer, G. H., M. B. Morrow, O. Wyss, T. E. Berg, and J. L. Littlepage. 1962. Antarctica: The microbiology of an unfrozen saline pond. *Science* **138**:1103–1104. The first description of the Don Juan Pond. The microbiology has not been confirmed.

Mooney, H. A., S. L. Gulmon, J. Ehleringer, and P. W. Rundel. 1980. Atmospheric water uptake by an Atacama Desert shrub. *Science* **209**:693–694.

Murray, B. C., M. E. Davies, and P. K. Eckman. 1967. Planetary contamination II: Soviet and U.S. practices and policies. *Science* **155**:1505–1511. See Horowitz et al. for part I.

Nobel, P. S. 1974. *Introduction to Biophysical Plant Physiology.* W. H. Freeman and Co., New York.

Robinson, R. A., and R. H. Stokes. 1968. *Electrolyte Solutions,* 2d ed. (rev.). Butterworths, London.

Rundel, P. W. 1982. Water uptake by organs other than roots. In O. L. Lange, P. S. Nobel, C. B. Osmond, and H. Ziegler, eds., Physiological Plant Ecology II, *Encyclopedia of Plant Physiology,* New Series, vol. 12B. Springer-Verlag, Berlin. A review that covers the uptake of water vapor by higher plants and lichens.

Sagan, C., E. C. Levinthal, and J. Lederberg. 1968. Contamination of Mars. *Science* **159**:1191–1196. A defense of the plan to heat-sterilize Mars landers.

Schmidt-Nielsen, K. 1964. *Desert Animals.* Oxford University Press.

Scott, W. J. 1957. Water relations of food spoilage microorganisms. *Adv. Food Res.* **7**:83–127. A historically important and still useful review of microbial water requirements.

Snyder, C. W. 1979. The planet Mars as seen at the end of the Viking mission. *J. Geophys. Res.* **84**:8487–8519. An excellent summary and bibliography.

Sverdrup, H. U., M. W. Johnson, and R. H. Fleming. 1946. *The Oceans.* Prentice-Hall, New York.

Vishniac, W. V., and S. E. Mainzer. 1973. Antarctica as a Martian model. In P. H. A. Sneath, ed., *Life Sciences and Space Research* XI:25–31. Akademie-Verlag, Berlin.

Weast, R. C., ed. 1974–1975. *Handbook of Chemistry and Physics.* CRC Press, Cleveland.

Chapter Seven

Biemann, K., J. Oro, P. Toulmin III, L. E. Orgel, A. O. Nier, D. M. Anderson, P. G. Simmonds, D. Flory, A. V. Diaz, D. R. Rushneck, J. E. Biller, and A. L. Lafleur. 1977. The search for organic substances and inorganic volatile compounds in the surface of Mars. *J. Geophys. Res.* **82**:4641–4658.

Chun, S. F.-S., K. D. Pang, J. A. Cutts, and J. M. Ajello. 1978. Photocatalytic oxidation of organic compounds on Mars. *Nature* **274**:875–876.

Clark, B. C., A. K. Baird, H. J. Rose, P. Toulmin III, R. P. Christian, W. C. Kelliher, A. J. Castro, C. D. Rowe, K. Keil, and G. R. Huss. 1977. The Viking X-ray fluorescence experiment: Analytical methods and early results. *J. Geophys. Res.* **82**:4577–4594. This paper describes the inorganic analysis of the surface of Mars.

Horowitz, N. H., G. L. Hobby, and J. S. Hubbard. 1976. The Viking carbon assimilation experiments: Interim report. *Science* **194**:1321–1322.

Horowitz, N. H., G. L. Hobby, and J. S. Hubbard. 1977. Viking on Mars: The carbon assimilation experiments. *J. Geophys. Res.* **82**:4659–4662.

Hubbard, J. S. 1979. Laboratory simulations of the pyrolytic release experiments: An interim report. *J. Mol. Evol.* **14**:211–221.

Hubbard, J. S., J. P. Hardy, and N. H. Horowitz. 1971. Photocatalytic production of organic compounds from CO and H_2O in a simulated Martian atmosphere. *Proc. Natl. Acad. Sci. USA* **68**:574–578.

Hunten, D. M. 1974. Aeronomy of the lower atmosphere of Mars. *Rev. Geophys. Space Phys.* **12**:529–535.

Hunten, D. M. 1979. Possible oxidant sources in the atmosphere and surface of Mars. *J. Mol. Evol.* **14**:71–78.

Klein, H. P., N. H. Horowitz, G. V. Levin, V. I. Oyama, J. Lederberg, A. Rich, J. S. Hubbard, G. L. Hobby, P. A. Straat, B. J. Berdahl, G. C. Carle, F. S. Brown, and R. D. Johnson. 1976. The Viking biological investigation: Preliminary results. *Science* **194**:99–105.

Levin, G. V., and P. A. Straat. 1977. Recent results from the Viking labelled release experiment on Mars. *J. Geophys. Res.* **82**:4663–4667.

Levinthal, E. C., K. L. Jones, P. Fox, and C. Sagan. 1977. Lander imaging as a detector of life on Mars. *J. Geophys. Res.* **82**:4468–4478.

Owen, T., K. Biemann, T. R. Rushneck, J. E. Biller, D. W. Howarth, and A. L. Lafleur. 1977. The composition of the atmosphere at the surface of Mars. *J. Geophys. Res.* **82**:4635–4639.

Oyama, V. I., and B. J. Berdahl. 1977. The Viking gas exchange experiment results from Chryse and Utopia surface samples. *J. Geophys. Res.* **82**:4669–4676.

Index